Social-Environmental Planning

The Design Interface Between Everyforest and Everycity

Social-Environmental Sustainability Series

Series Editor

Chris Maser

Published Titles

Social-Environmental Planning: The Design Interface Between Everyforest and Everycity, **Chris Maser**

Forthcoming Titles

Sustainable Development: Principles, Frameworks, and Case Studies, **Okechukwu Ukaga, Chris Maser, and Michael Reichenbach**

Social-Environmental Planning

The Design Interface Between Everyforest and Everycity

Chris Maser

CRC Press
Taylor & Francis Group
Boca Raton London New York

CRC Press is an imprint of the
Taylor & Francis Group, an **informa** business

CRC Press
Taylor & Francis Group
6000 Broken Sound Parkway NW, Suite 300
Boca Raton, FL 33487-2742

Printed in the United States of America on acid-free paper
10 9 8 7 6 5 4 3 2 1

International Standard Book Number: 978-1-4398-1459-8 (Hardback)

Library of Congress Cataloging-in-Publication Data

Maser, Chris.
Social-environmental planning : the design interface between everyforest and everycity / Chris Maser.
p. cm. -- (Social environmental sustainability)
Includes bibliographical references and index.
ISBN 978-1-4398-1459-8 (alk. paper)
1. Sustainable development--Social aspects. 2. Environmental policy--Social aspects. 3. Environmental protection--Planning. I. Title. II. Series.

HC79.E5M35974 2010
306.3--dc22 2009010929

Visit the Taylor & Francis Web site at
http://www.taylorandfrancis.com

and the CRC Press Web site at
http://www.crcpress.com

Contents

Section II Nature's Rules of Engagement in Social-Environmental Planning

Section IV A Century for Healing

Acknowledgments

It is with great pleasure that I thank Jane Silberstein (Cooperative Extension, University of Wisconsin, Ashland, and an independent city planner) and Okechukwu Ukaga (Executive Director, Northeast Minnesota Sustainable Development Partnership, University of Minnesota, Duluth) for taking their precious time to critically review and greatly improve this book.

As always, Zane, my lovely wife, deserves special appreciation for her patience with the many hours I spent working on the manuscript, which deprived her of things we might have done together. I am also grateful for her review of Chapter 3. Thank you.

Finally, it has been a genuine pleasure to work with Irma Shagla, Acquisitions Editor, as well as the other thoughtful professionals at CRC Press/Taylor & Francis Group. As well, I am deeply grateful to Kathleen Brown for an excellent job of copyediting.

Introduction

Wherever I have been, from Oregon to Alaska and Canada, from Egypt to Nepal, Japan, Eastern Europe, Malaysia, Chile, and so on, I have seen the similarities of Nature. And yet, as a scientist, I was trained to focus on Nature's differences, which I suppose should come as no surprise since our American culture is primarily a divisive one that tends to focus on our differences as a collective people rather than on our similarities. In some respects, we are still a nation of "settlers" from many different cultures, staking our own claims and competing for resources through the "money chase."

A simple way to envision the idea of commonalties might be a Shakespearean play. While the play is the same over and over and over, the stage, props, and actors are changeable and can be different each time the play is enacted but without altering the theme or the language the play expresses. As Shakespeare's English is the language of the play regardless of when, where, or by whom it is performed, so Nature's biophysical principles embodied in cause and effect are the language of all viable elements of design, be it Nature's or humanity's. To make this thought more concrete, I offer a personal experience. Although I have used this analogy elsewhere, I know of none more dramatic in the clarity of its interface between design and systems function.

While I was working in Nepal some years ago, a helicopter crashed and killed two people. A helicopter, as you might imagine, has a great variety of pieces with a wide range of shapes and sizes, of which the mechanic responsible for the helicopter's maintenance knows the individual arrangements and functions. The problem with this particular helicopter was in the engine, which was held together by many nuts and bolts. Each nut and bolt has a small, sideways hole drilled through it so a tiny "safety wire" can be inserted. The ends are then twisted together to prevent the tremendous vibration created by a running engine from loosening the nut, thus allowing it to work itself off its bolt. Simply put, the function of the small hole and little piece of wire is to counteract the engine's vibration.

Prior to its last maintenance, the helicopter had functioned as it was designed to and had remained safely airborne for many hours. At the required time, it was grounded for maintenance. On its first postmaintenance flight, however, it crashed into the jungle without warning. Why? On retrieval of the helicopter, mechanics spent many hours examining all of its pieces to see what had gone wrong. At length, they found out.

One of the mechanics who had helped perform the helicopter's last maintenance had forgotten to replace one tiny, four-inch piece of safety wire that held a nut in place on its bolt, which in turn kept the lateral control assembly together. The nut had vibrated off its bolt, the helicopter had lost its stability, and the pilot had lost control. All this was caused by one missing piece of wire that altered the entire functional dynamics of the aircraft. The engine had been "simplified" by the lack of a single, out-of-sight component.

Which piece—at that critical instant—was the most important part in the helicopter? Which element of design in a village, town, or city is the most important? Which component in the design of a landscape or a forest is the most important?

Clearly, each part of a system has a corresponding relationship with every other part, and they provide stability only by working in concert within the limits of their designed purpose. Ultimately, the elements of design are basically the same in both a city envisioned by humans and a landscape created by Nature. That is, all of the design elements humanity incorporates into the building and maintenance of a human community of any size are inexorably based on the design elements found in Nature, for example:

- Velcro is an idea borrowed from the hooks and grabbing function of cockleburs.
- The manufacture of a light, but tough, police flack jacket was inspired by the composition and strength of a spider's web.
- A dam in a river was designed first by beavers.
- A subway system is patterned after the subterranean burrow system of a mole or gopher.
- An apartment complex is a structure with a variety of chambers committed to different uses by a number of unrelated individuals; as such, an apartment complex shares design elements found in termite colonies of the Australian savannah (many chambers of various sizes and uses found in a single structure) and in the adjoining nests of North American cliff swallows (shared walls that separate chambers occupied by unrelated individuals) and some African weaver finches (a few actually accommodate more than one couple per chamber in bulky, woven nests).
- A housing development emulates a colony of North American bank swallows and Nepalese sand martins, both of which have nests of unrelated individuals clustered in a common area along the high bank of a river but not immediately adjoining one another.
- The manufacture of synthetic pharmaceutical products is based on decoding how chemical compounds found in medicinal plants used by indigenous peoples function to cure aliments and then replicating them.

While the design elements humanity has appropriated from Nature for its own purposes seem infinite at first glance, they are surprisingly repetitive when considered within the biophysical constraints that govern our home planet and its place in the universe. In this book, I present some of the myriad commonalties of design between a human community and Nature and provide examples of actual design interfaces from a community perspective. Of necessity, I must deal first and foremost with the geographical area and ecosystems wherein I have done the most research and thus understand the best. Nevertheless, while some of the plants and animals I cite as examples are from my geographical area of expertise, the principles and processes described within these covers are global.

The use of "Everyforest" and "Everycity" in the title and text of this book is patterned after the short, 900-line, Flemish play first printed in 1495. The play depicts a complacent Everyman informed by Death of his impending demise. Everyman dramatizes the universal struggle of every individual. In essence, the play is a reminder that we can take with us into death only what we have given, such as "Good Deeds," but nothing of a material nature.

I dare to undertake this book fully acknowledging the compounding uncertainties humanity is facing, a few of which are a soaring human population; continual loss of biological, genetic, and functional diversity; growing pollution of air, water, and soil; depletion of the ozone layer; a changing global climate; crumbling families; and the loss of trust among community members. I am also aware that humanity's general response to these challenges is *symptomatic* as opposed to *systemic*, a mindset that is like visiting a doctor to cure a symptom but refusing to change one's lifestyle—the cause of the symptom.

With the above in mind, I see two simultaneous approaches humanity must take if social-environmental sustainability has any chance of becoming a legacy we adults can bequeath the children of any generation, let alone all generations. These approaches are to (1) learn to adapt to the circumstances of global climate change as they already are and (2) determine what we can do to stabilize the climate as effectively and quickly as possible. There is a necessary caveat here; namely, change is a constant process by which eternal novelty and everlasting irreversibility are duel characteristics of the biophysical principle of *cause and effect*, a principle that governs our lives first and foremost.

Therefore, our *willingness* to alter our behavior in response to *inevitable* change will indeed determine how long today's communities—and society itself—can survive. None of these problems can be mended with the current scientific Band-Aids, technological quick fixes, or political rhetoric—no matter how good it sounds or comforting it might be. Nevertheless, out of the current social chaos can arise a society better balanced among the scientific and the social, the materialistic and the spiritual, the masculine and the feminine, the intellectual and the intuitive, the present and the future, and the local and the global. To achieve this better balance necessitates that we each find the personal courage and political will to take responsibility for our own behavior and thereby elevate our consciousness of cause and effect to view the world and society with a more holistic frame of reference.

Achieving such an elevated view will be difficult because humanity has long taken the world designed by Nature for granted and has exploited it to such an extent that human society cannot long endure with any sense of well-being and dignity on its present course. People within a community compete with one another for the goods and services of Nature—depleting our common natural wealth with no thought of reinvestment in its potential sustainability. In turn, each community competes with every other community within a society, and each society competes with every other society for the same goods and services. In that competition, each community, and thus society itself, has become so needy and so specialized in the materialistic sense that today we live in a global collection of competing societies standing precariously like a house of cards. If one society falls, the ripples of collapse are felt throughout the world, at times with stunning rapidity and progressively irreversible consequences.

Here, it is crucial to understand that, social-environmental sustainability is a *process* and not an endpoint. A process is the functional interaction of two or more components that formulate a necessary sequence of steps toward an outcome—the product. But first, we need to know what outcome we want. Then, we need to focus on the processes necessary to achieve that outcome with a consistent, persistent, and disciplined effort. If we focus in error only on the outcome, we destroy the process, as clearly stated in Thomas Merton's translation of Chuang Tzu's classical poem:

> When an archer is shooting for nothing he has all his skills.
> If he shoots for a brass buckle, he is already nervous …
> The prize divides him.
> He cares.
> He thinks more of winning than of shooting—and the need to win
> Drains him of power.[1]

If humanity is to grace this marvelous planet Earth with any degree of honor and dignity, the day has to arrive when we, its citizens, come to understand just how imperative it is that we set aside our historic environmental, ideological, and religious competition and begin instead to cooperate with one another for the common good—present and future. Only then will we be able to bring our various cultures into social-environmental harmony so there will be room for all planetary citizens, both human and nonhuman. Only then are changes wrought by the human hand likely to be caring, with respect to this grand experiment called "life." Nevertheless, despite our best intentions, despite our greatest knowledge, social-environmental sustainability is a process for which continual maintenance and repair are fundamental necessities.

In this book, I weave the shared threads of design between a forest and a human community into a conceptual framework that depicts their commonalties and demonstrates how we can design communities in a way that consciously supports the beauty and functionality of Nature's design. If you are wondering why I have chosen a forest to compare with a city, the reason is twofold: (1) Humanity's primary source of water—for which there is no substitute—comes mainly from forested water catchments worldwide, and (2) a forest of some kind not only is the most commonly visited ecosystem on Earth but also has the most easily discernable components, despite being the most complex of all terrestrial bionetworks.[2]

In this sense, a forest (and in some respects, an old forest) is the classroom for learning about Nature's biophysical principles as functional elements of design. The city, on the other hand, is the practicum wherein we learn how to apply the biophysical principles, and through their application, to test our understanding of them.

In depicting Everyforest and Everycity, the superficial appearance of each is seen as vastly different, whereas the dynamics that create and maintain them are basically the same:

- In human endeavors, for example, a society functions as a whole within a specific "frame of reference" and thereby serves to identify each society within itself, while differentiating it from others sharing the same space-time continuum. In Nature, the inviolate principles governing all biophysical processes act as the integrating frame of reference that simultaneously binds together the universal commonalities while allowing the novelty of differences.
- In our social-environmental world, the biophysical principles encompass the ever-widening circles of cause and effect and so act as the governing strands that give form and function to Nature—including humanity as an inseparable part thereof.
- In human endeavors, our decisions and subsequent actions result in substantive outcomes, each constrained by the same principles embodied in cause and effect that govern Nature's biophysical processes.
- Nature's colors and hues are displayed in flower, butterfly, and autumn leaf, in sea and sky. In humanity's realm, colors and hues are conveyed through art and language.

Every human language—the master tool representing its own culture—has a unique construct, which determines both its limitations and its possibilities in expressing myth, emotion, ideas, desires, and logic. As such, language is the medium through which the spiritual condition of the human soul is painted and the material condition of human desire is wrought on the land.

We humans compose the broad shapes of a cultural story line by using words to convey colors of meaning we mix on a palette of syntax. Then, by matching the colors of words to grant expression to our ideas, we add verbal structure, texture, shades of meaning, and hue to the story. In doing so, we create a picture or portrait as fine as any achieved with brush, paint, palette, and canvas; with camera and film; or with musical instruments and mute notes on paper. In addition, a verbal picture often outlasts the ravages of time that claim those pieces of art on canvas, imprints of light on photographic paper, or instruments that give "voice" to mute shapes.

- Design is the coming together of texture, color, hue, and form that bestows visible expression to something heretofore unseen. Nature's design is consummated through the biophysical attributes and processes that ultimately constitute life, whereas humanity's design culminates in the material manifestation of an inner desire conveyed through the medium of language.
- In Nature, we see the ever-changing flow and ebb of process and novelty. Our intellectual understanding of Nature, both scientific and intuitive, allows us to pattern processes and products after Nature in attempting to bring forth social-environmental sustainability.

Finally, my purpose in writing this book is to help people accept what is by understanding the commonalities of pattern between Everyforest and Everycity. With this comprehension, people can adapt to the present and begin determining what they can do now to leave this magnificent planet a little better for each succeeding generation. After all, the Earth is a biological living trust, and we, the adults, are the trustees for all future generations.

To better effect our trusteeship of our home planet, I have divided this book into five parts: Part I deals with the purely human dimension of our social-environmental landscape and is not about the elements of design per se but rather about the way we *perceive* those elements and how we share our perceptions. Understanding how we communicate our perceptions with one another is critical if we are to honor the parameters of Nature's biophysical principles and thus maintain our ecosystems intact while we plan our villages, towns, and cities—and so redesign our shared landscape.

Part II is a brief treatise of Nature's rules of engagement in life and human behavior, both of which pertain to social-environmental planning. To this end, Chapter 3 examines the inviolate biophysical principles that regulate how our world functions.

Part III is a comparison of the design elements common to Everyforest and Everycity; wherein is discussed how Nature's biophysical principles and the dynamics of our behavior intersect. This discussion is necessary because somewhere in our travels through time we have lost sight of the fact that we all live and die according to Nature's rules as they pertain to social-environmental sustainability. For this reason, we have to understand the biophysical dynamics of Nature to understand the cause-and-effect relationships set in motion by our urban designs. In turn, we have to understand the cause-and-effect relationships set in motion by our urban designs to understand and protect the productive capacity of our surrounding landscape on which we and future generations must rely for products and amenities, such as potable water—the essential fluid of life.

Part IV is an exploration of the reciprocal relationships (midwifed by the common elements of design) between Everyforest and Everycity.

Finally, the gist of Part V is the determination that an elevation of consciousness is necessary if we are to heal ourselves and thereby ensure the reciprocal integrity of the relationships among our villages, towns, and cities and the landscapes that sustain them—present *and* future.

Notes

1. Thomas Merton. *The Way of Chuang Tzu*. New Directions, New York (1965).
2. For an in-depth discussion of the global commonalities of forest dynamics, see Chris Maser, Andrew W. Claridge, and James M. Trappe. *Trees, Truffles, and Beasts: How Forests Function*. Rutgers University Press, New Brunswick, NJ (2008).

About the Author

Chris Maser has more than 40 years of experience in ecological research, including broad international experience; he has spent many years developing new ways of understanding how forests are structured and function as well as new ways of understanding social-environmental sustainability and conflict resolution. He has written over 250 articles (mostly in scientific journals) and some 29 books on these topics. His books, as well as some of his articles, are in academic and public libraries in every state in the United States and all but one province in Canada, as well as in 70 other countries. He has lived, worked, or lectured in Austria, Canada, Chile, Egypt, France, Germany, Japan, Malaysia, Nepal, Slovakia, Switzerland, and much of the United States. His most recent book, *Earth in Our Care: Ecology, Economy, and Sustainability,* was published in 2009 by Rutgers University Press. (See his Web site if you want more information: www.chrismaser.com.)

Section I

The Human Dimension of Social-Environmental Planning

Our understanding of the world, both scientific and intuitive, serves as the framework for our decisions and subsequent actions as they shape a landscape of social-environmental design based on our sense of values, but within the constraints of Nature's biophysical principles.

1

How We Think

Introduction

Every child is born with a personal "scorecard"—a blank slate to be filled in as life progresses. But, no child is initially allowed to keep score because parents with their "dos" and "don'ts," "shoulds" and "shouldn'ts"—mostly the latter, almost immediately usurp that right. And then, of course, there is "good" and "bad."

Although a child's parents are the initial scorekeepers, that activity soon becomes the shared property of peers, schoolteachers, university professors, and society at large. The challenge faced by today's children is the unfortunate circumstance of being continually taught in the negative. They are continually encouraged to move *away* from what their parents, teachers, and peers do not want rather than deciding for themselves what they *do want* and moving toward it.[1]

Consequently, a person's *frame of mind* is the amalgam of that person's upbringing expressed as the breadth of his or her thinking and the rigidity of personal intellectual boundaries, which of course affect the person's view of the world. Concomitantly, a *frame of reference* is the shared understanding of individuals regarding their experiences, concepts, values, customs, and perceptions.

Frame of Mind

Each person's frame of mind both adds to and compounds social diversity not only because perceptions differ at any point along the continuum of life but also because each is influenced by age and the accrual of life's experiences. A person's frame of mind also varies with the person's degree of focus, centeredness, personal identity, formal education (both secular and religious), and degree of spirituality. Moreover, most people react to social pressures and the ever-shifting relationships of human beings to one another, each of whom has a different worldview, one that is constantly changing, if only slightly. Consequently, how we each understand and accept "what is" ultimately determines how we live life, and that in turn is expressed as a set of general personality traits.

Our Shared Personality Traits

In a sense, generalized personality traits are an amalgamation of the dominant ways in which each person navigates life. They emerge from an interpretation of life's experiences and are the springboard of personal capabilities. These traits are not cut-and-dried but rather overlapping tendencies with varying shades of gray. While in the collective they form the diversity necessary for the existence of a viable human community, they can also be substantial barriers to sharing life's experiences and thus a common understanding of the world in which we live—the world we are constantly redesigning.

For instance, some people can take ideas seemingly at *random* from any part of a thought system and integrate them; these people have mental processes that instantly change direction, arriving at the desired destination in a nonlinear, intuitive fashion. Others tend toward thinking in a *linear sequence*, like the coupled cars of a train, wherein the mental processes crawl along, exploring this avenue and that, without assurance of ever reaching a definite conclusion. If the random thinker is at ease with *abstractions* but the linear, sequential thinker requires *concrete* examples of what is being discussed, their attempts to communicate may be difficult.

These two approaches can be thought of as *product-oriented thinking* and *systems-oriented thinking*. Product-oriented thinkers tend to focus on perceived products—economically desirable pieces of a system—in isolation of the system itself. Thus, they are likely reticent to accept the fact that removing either a desirable or an undesirable piece of a system can or will negatively affect its productive capacity as a whole. Such a restricted view usually breeds a misunderstanding of both the component and the system of which it is an inseparable part. System-oriented thinkers, on the other hand, are inclined more toward a holistic approach by focusing primarily on the processes that govern a system's functional capacity.

To demonstrate how these two ways of thinking might play out in the public arena, imagine a couple of scenarios within a city to which the council and its planning staff must respond. In the first scenario, a small group of people living in the outskirts of the city want to build a church whose membership would most likely have a maximum of fifty or sixty members over time. In the second scenario, a developer wants to build a shopping mall in a densely populated suburb well within the city limits. In both cases, an issue arises with the parking lot. First, let us deal with the church.

When the church members present their plan to the city council and the planning staff, it becomes apparent that cost is an issue. To alleviate some of the potential debt burden, the church members want permission to gravel their parking lot instead of having it covered with asphalt. On hearing this request, a member of the planning staff immediately says that it is impossible to gravel a parking lot because the regulations call for the surface of all parking lots to be asphalt. On this point, the staff member is adamant.

At length, after much discussion and the continual staff rejection of the church's proposal, the church members say they would appeal the decision. There was, however, one member of the city council who had listened quietly to the whole discussion. Only now, when the church members say that they will appeal, does the councilperson speak.

To everyone's surprise, the councilperson points out that a graveled parking lot makes good sense in the case of the church for the following reasons:

1. With a graveled lot, the infiltration of rain would help recharge the groundwater, which was acceptable in this case because the parking lot was far enough removed from the aquifer the city used for its drinking water that the lot could not possibly affect the water's quality.
2. There would be relatively few vehicles parked on the lot at any one time, and then infrequently, so pollution from oil and the like would be minimal.
3. There would be no need for a connection to the city's storm-drain system because the infiltrating water would be purified by its slow travel through the soil toward the distant river.
4. The city would save money over time on the inevitable maintenance of its storm-drain system because an extension would not have to be installed.
5. Additional money would be saved because the water would infiltrate into the soil, rather than being collected in the storm-drain system, where it would pass through the city's water-treatment plant, adding to the annual cost of the plant's operation.
6. As well, the surface of the parking lot would be much less expensive to maintain if it was graveled instead of paved.

After more debate, the city council voted to extend a waiver and allow the church to gravel its parking lot.

Hearing about the waiver granted the church, the shopping mall developer goes to the city council and the planning staff and requests the same waiver for his shopping mall, putting forth the argument that it would save the city money. But in this case, the vote is unanimous in opposition to the waiver, which is denied because of the following:

1. The aquifer from which the city draws its drinking water flows directly under the site of the prospective shopping mall.

2. The volume of vehicles would discharge so much pollution over time that the probability of its negatively affecting the aquifer is virtually certain.
3. It is in the long-term interest of citizens to protect the quality of the water from this source of potential pollution.
4. If the aquifer becomes polluted, it would affect the entire city as an irreversible, negative circumstance.
5. The immediate cost to the developer of paving the parking lot would be negligible when compared to the inevitable, long-term, negative, social-environmental impact of such a waiver and the ultimate cost to the citizens in a reduced quality of life.

In this example, the case is successfully made for a systems approach to a land-use decision, suggesting that a person's understanding can change if the matter is presented in a logical, dignified manner. We can surmise, therefore, that the scope of a person's frame of mind, which simultaneously represents the person's own familial and cultural foundation expressed as conceptual limitations, has its unique construct, thereby determining the possibilities of the person's understanding.

These traits, coming in a variety of combinations, indicate how different and complex people can be in response to their life experiences. When people's experiences and their shared understanding meld, a predominant worldview emerges—one tempered by each person's degree of self-control.

Our Thoughts, Motives, and Actions Are All We Can Ever Control

Bluntly put, our thoughts, motives, and actions are all we can ever control despite how fervently we might wish it otherwise in our vain attempts to *manage* our environment. The fact is, we cannot even manage ourselves all that well, witness the growing number of addictions that plague industrial society (notably in the United States): nicotine,[2] alcohol,[3] drugs,[4] gambling,[5] and junk food.[6] This says nothing about the growing obesity,[7] stress, fear, and greed that seem to have become increasingly pervasive in the first decade of this century. If we are unable to manage ourselves, how can we manage anything outside ourselves?

We introduce thoughts, practices, substances, and technologies into the environment, and we usually think of these introductions in terms of *development*—primarily economic development. Development of any kind is the collective introduction of thoughts, practices, substances, and technologies into a commercial strategy to access a given resource through our notion of *management*. Whatever we introduce into the environment in the name of development will consequently determine how the environment will respond to our presence and to our cultural necessities. So, it is to our social benefit to pay close attention to what we introduce because these things represent both our sense of values and our behavior.

Our initial introduction is our pattern of thought that determines the way we perceive the Earth and how we act toward it—either as something sacred to be nurtured or as only a commodity to be converted into money. Because our pattern of thought determines the value we place on various components of an ecosystem, it is our sense of values that determines the way we treat those components and through them the ecosystem as a whole.

In our predominantly linear, product-oriented thinking, any natural resource is an economic waste if its "conversion potential" is not realized—that is, the only value a resource has is its potential for being converted into money. Such notions stimulated Professor Garrett Hardin to observe that "economics, the handmaiden of business, is daily concerned with 'discounting the future,' a mathematical operation, that under high rates of interest, has the effect of making the future beyond a very few years essentially disappear from rational calculation."[8] Unfortunately, Hardin is correct. Conversion potential of resources counts so heavily because the effective horizon in most economic planning is only five years away. Thus, in our traditional linear economic thinking, any resource that is left *unused* by humans is thought to be an *economic waste* because its potential as an economic product was not realized, even if, like a tree lying on a forest floor, it reinvests its biological capital (fertility) into the soil.

The initial impetus to pollute our common environment, such as air, water, and soil, is caused by our thinking as played out in our myopic focus on the money chase. And, lest you think otherwise, the moment we introduce something into our global environment, it is immediately out of our control, such as medicines, vitamins, and health supplements of one kind or another that pass through our bodies and get flushed down the toilet. Yet, the effluent from the manufacture of pharmaceuticals is often worse.

The drug manufacturers in Patancheru, near Hyderabad, in southern India, serve as illustrative of the environmental pollution caused by a single, major production site of generic drugs for the world market. In Hyderabad, the industrial plant that processes effluent from the ninety large pharmaceutical manufacturers in Patancheru discharges the highly contaminated water into a stream that eventually joins the Godavari River, the second largest river on the subcontinent emptying into the Indian Ocean. Once in the ocean, it contaminates all the connected water on Earth through a system of shared marine currents.

The released water contains astronomical amounts of antibiotics, along with large concentrations of analgesics, drugs for hypertension, and antidepressants. What is more, in keeping with a common practice, the treatment plant mixes raw human sewage with contaminated effluent, which contains enormous quantities of antibiotics that will encourage the evolution of bacteria to resist these same antibiotics.

Ultimately, pharmaceuticals, ranging from painkillers to synthetic estrogens, are entering the waterways of the world, and thus our common environment, through human excreta, hospital and household wastes, and agricultural runoff, as well as from water treatment plants. Synthetic estrogens and their mimics are known to have negative impacts on the sustainability of populations of wild indigenous fish as well as on the developmental processes of amphibians in streams that receive polluted water from municipal wastewater treatment plants.[9]

Another way medical waste is increasingly spread throughout various parts of the environment is by the distribution of sewage sludge as fertilizer.[10] Most of this material, along with a vast array of other chemicals, ends up polluting the groundwater.[11] "Once polluted," warned ecologist Eugene Odum, "groundwater is difficult, if not impossible, to clean up, since it contains few decomposing microbes and is not exposed to sunlight, strong water flow, or any of the other natural purification processes that cleanse surface water."[12]

Knowledge Is Some Version of the Truth

The realities we accept as obvious, neutral, objective, and simply the way the world works are actually structures of power we created as we think and live. They are created by our rendition of history, our understanding of ourselves, of our society, and of our world—and they are always partial with respect to the whole.

Over the years I labored as a research scientist, I came to appreciate how much—and yet how *very, very little*—we humans understand about ourselves, let alone the world of which we are an inseparable part. There is so much for us to learn about ourselves as individuals, as a species, and about the Earth we influence in our living, that I firmly believe the complexities of life and its living are permanently beyond our comprehension. Because knowledge is always relative, we can only navigate an ever-shifting version of the truth through the continuous accrual of knowledge. This being the case, the salient point is not our knowledge but rather our ignorance because only our ignorance can be proven, as Albert Einstein noted when he said, "No amount of experimentation can ever prove me right; a single experiment can prove me wrong."[13] Thus, the validity of our knowledge, like all our versions of the truth, rests, albeit tenuously, before the jury of tomorrow, the day after that, and the day after that, ad infinitum.

Furthermore, we are subjective beings who cannot have an objective (neutral) thought if for no other reason than because every person sees through their own lens but dimly. (If you doubt what I just said, try asking a *neutral* question.) Such is the case because we cannot detach ourselves from Nature. In addition, all we can judge as fact is our own perceptions, always colored as they are by our personal lenses. Even history, which we tend to think of as fact, is viewed individually through interpretation of what we perceive.

The irony in our search for knowledge is that nothing can be proven—only disproved. Therefore, we can never know *the* truth of anything in terms of knowledge because it is explainable only in the illusions of its appearance, something that may have prompted Sir Francis Bacon to say, "Science is but an image of the truth."[14] Thus, the actual objects of our inquiries, the formulations of our questions and definitions, and the mythic structures of our theories and facts are social constructs. Every aspect of our theories, facts, and practices—including the methodology by which they are derived—are but expressions of contemporary social, political, and economic interests, cultural themes and metaphors, personal biases, and personal/professional negotiations for the power to control the knowledge of the world, albeit momentarily and minutely.

The problem is that we confuse the limiting nature of scientific inquiry with the nature of ordinary experience. While scientific method tends to be reductionistic and mechanical, life experiences are holograms made of complex threads of past experiences, present perceptions, dreams of future opportunities, and fears of future disasters. Life is thus a moment-by-moment kaleidoscope of a present moment in eternal flux within and between internal and external renditions of perceived realities and may or may not have anything consciously in common with *reality as reality*. This simply means that something can be proven only by its actual occurrence, meaning "proof" comes after the fact—not before it.

Although Nature provides a degree of predictability over time through the ability to recognize and read a trend, we remain stubbornly committed to the concept of an absolute. The concept of an absolute probably arose in response to human fear of unknown but observable natural forces. This fear may have given rise to religious ideas conceived in "the necessity of defending oneself against the crushingly superior force of nature."[15]

In the sixth century BCE, the roots of Western science arose in the first period of Greek philosophy, a culture in which science, philosophy, and religion were united. A split in this unity began with the assumption of a Divine Principle that stands above all gods and people. Thus began a trend of thought that ultimately led to the separation of spirit and matter and to the dualism that characterizes Western philosophy.[16] Because the Christian church supported Aristotle's view that questions concerning the human soul and God's perfection took precedence over investigations of the material world, Western science did not develop further until the Renaissance, when Sir Francis Bacon in the late 1500s gave humanity the ability to experiment and defined what is now called the scientific method. "Look," he told the world, "There is tomorrow. Take it with charity lest it destroy you."[17]

Thus, separation of knowledge and intuition began with the ancient Greeks. Western philosophy embraced what is thought to be rational knowledge: that derived from experience with objects and events in one's immediate environment. Such knowledge belongs to that realm of the intellect that discriminates, divides, compares, measures, and categorizes. It creates a world of apparent distinctions whose opposites can exist only in relation to one other.

Abstraction, according to physicist Fritjof Capra, is a crucial feature of knowledge because in order to compare and classify the immense variety of sizes, shapes, structures, and phenomena that surround us, we must select a few significant features to represent the incomprehensible milieu. We thus construct an intellectual map of reality that reduces things to their general outlines, an action resulting in knowledge being a system of abstract concepts and symbols characterized by the linear, sequential structure that is typical of our thinking and speaking.[18]

The natural world, on the other hand, is one of infinite variables and complexities, a multidimensional world without straight lines or completely regular shapes, where things do not happen in sequences, but all at once, a world (as modern physics shows us) where even the void of space is curved. It is thus clear that our abstract system of conceptual thinking is incapable of completely describing or understanding this reality and so is necessarily limited.

If, therefore, we are going to ask intelligent questions about the future of the Earth and our place in the scheme of things, we must understand and accept that most of the questions we ask deal with cultural values and cannot be answered through scientific investigation. Nevertheless, scientific inquiry can help elucidate the outcome of decisions based on these values and must be so employed. We must also be free of scientific opinions based on any special interest group's "acceptable" interpretations of knowledge. In addition, we would be wise to accept the gift of Zen and approach life with a beginner's mind as our frame of reference—a mind simply open to the innovative possibilities for maintaining the social-environmental sustainability of planet Earth.

Frame of Reference

The social-environmental landscape, of which we humans are an integral and inseparable part, is composed of and affected by the collective thoughts and attitudes of the citizenry and so constitutes all levels of the body politic. Our frame of reference consists of an apparent tripartition: perception, paradox, and unity.

Perception

Perception, from the Latin *percipere* (meaning to see wholly), is a composite of one's past experiences, present sense of reality, and expectations in the future, precluding any two people from seeing the same thing in the same way. Of perception, poet William Blake wrote: "If the doors of perception were cleansed everything would appear as it is, infinite."[19] Instead of "doors," think of perception as a "window" through which we peer at the world around us—a window that often becomes increasingly difficult to see through with any semblance of clarity as we mature into adulthood.

Although the window of adult perception is increasingly smudged with fears, the window of childhood perception is crystalline in its transparency. It is a precious gift we lose as our childhood innocence is progressively stolen or instantly dashed by adults who often see but dimly through their fears—adults who perceive the world to be such a dangerous place they have lost sight of its magnificence.

Under the tutelage of adults, children are taught to disregard, distrust, or even deny "unacceptable" aspects of their perception. In this way, children are taught which "truth" is acceptable and which is not.

Because each person's perception is that person's truth, each of us is right from our point of view, whatever it is. In other words, you are right from your point of view, whatever it may be, and so is everyone else. We can, if we choose, view the world and one another from the position of "I'm right *and* you're right" (albeit different) in our respective points of view—as opposed to "I'm right so *you have to be wrong*." With the former view, negotiating a new relationship with one another and with our home planet would be much easier than constantly fighting the emotional gridlock we all too often find ourselves in because of our insistence on the irreconcilable exclusivity of *either/or* thinking instead of accepting the inclusivity of *and* thinking.

The Indian spiritual leader Mahatma Gandhi said, "A votary of truth is often obliged to grope in the dark."[20] Our challenge, as Gandhi pointed out, lies in our blind spots, not in our vision. Unlike correcting a blind spot in the rear view of an automobile simply by adding a different kind of mirror or a supplemental one, we cannot rectify our personal blind spots so easily. To correct them necessitates personal growth in both our perception and our acceptance of what is.

For Socrates, true knowledge was more than a simple inspection of so-called facts, although these facts can provide important information. Rather, true knowledge is the power of the mind to understand the enduring elements of cause and effect that remain after the illusion of facts disappears. While our perceptions grow and change as we mature, not everyone's perceptions mature at the same rate, which accounts for the widely differing degrees of consciousness with respect to cause-and-effect relationships. This disparity is neither good nor bad; it simply means that each of us has different gifts to give at different times in our life as we perceive truth differently.

Truth is *absolute*; perceptions of truth are relative, which means facts, as perceptions of truth, are relative. Think carefully about the following statement: The world functions perfectly; our perception of how the world functions is imperfect. We assume this statement to be true because it accepts Nature's biophysical principles as absolute truth. But what are those principles? How do they work? We do not know for sure because our perception is constantly changing as we increase the scope of our knowledge.

Trying to understand Nature's biophysical principles is the essence of science. Yet, a fact is a fact by one of two means: (1) by direct observation (my neighbor's tree fell on their house during the storm yesterday) and (2) by consensus of scientists who desist in attempting disproof, which means that a scientific fact or truth is merely an approximation of what is. It represents our best understanding of reality at a particular moment and is constantly subject to change as we learn.

Perception *is* learning because cause and effect are always connected. Mahatma Gandhi reached this conclusion when he said:

> At the time of writing I never think of what I have said before. My aim is not to be consistent with my previous statements on a given question, but to be consistent with truth as it may present itself to me at a given moment. The result has been that I have grown from truth to truth; I have saved my memory an undue strain; and what is more, whenever I have been obliged to compare my writing even of fifty years ago with the latest, I have discovered no inconsistency between the two.[21]

Gandhi was consistent in his changing perceptions of what "the truth" was at different stages in his life. He grew from truth to truth as his vision cleared and he could see greater vistas. He said that if one found an "inconsistency" between any two things he wrote, the reader "would do well to choose the latter of the two on the same subject."[22]

The accepted definitions of truth are only modifications of the definitions of perception. Truth as a human understanding resides in everyone's heart, and it is there one must search for it. Although we must each be guided by truth as we see it, no one has a right to coerce others to act according to his or her view of truth. In the end, our "detector of truth" is our *inner voice*. Ergo, there is no magic in the perfection of hindsight; it only points out that we did not listen to our inner voice when it spoke the first time.

The truth of the human mind is relative and so merely a perception of what is true. If our perception of a truth were in fact *the truth*, we would find knowledge to be absolute and no such thing as a *version* of the truth. Truth is perfect understanding of what is. It is neither the spoken nor the written word, although these may have a *ring of truth* about them.

Nevertheless, each person's frame of reference both adds to and compounds diversity not only because our perceptions are different from one another at any given point along the continuum of life but also because they change with age and the accrual of life's experiences. In this sense, each human being is the sum of his or her perceptions of life. So it is that human perception at once creates, integrates, and re-creates itself in an ever-widening sphere of consciousness. Because we manifest what we think, the richness or poverty we create through our life experiences, individually and together, is our choice.

A person who tends to be positive or optimistic sees the proverbial glass of water as half full, whereas a person inclined to be negative or pessimistic sees the same glass as half empty. Regardless of the way it is perceived, the level of water is the same, illustrating that we *choose* to interpret what we see, and that may have little to do with reality.

Therefore, if we can agree that everyone is right from his or her point of view, and that each point of view is only a different perception along the same continuum, then it would seem reasonable that the freer we are as individuals to change our perceptions—without social resistance in the form of ridicule or shame—the freer society is to adapt to change in a healthy, evolutionary way. On the other hand, the more rigidly monitored and controlled "acceptable" perceptions are (i.e., politically), the more prone a society is to the cracking of its moral foundation and the crumbling of its knowledge-based infrastructure because nothing can be held long in abeyance in an ever-evolving world, least of all social evolution.

I say *social* evolution because humanity constantly alters its culture through a process of trial and error—"mutations" of heretofore accepted social behavior, much the same as genetic mutations are interjected into a species' lineage and mediated by the biophysical principles of survival in a particular environment. As such, intrinsic social elements are encompassed in ideas; through time, these ideas become expressed as changing social values, causing extrinsic changes in human behavior that ultimately affect the social-environmental condition bequeathed to future generations. In this way, each perception, representing as it does an individual's personal and cultural foundation, has its unique construct and determines the possibilities of the individual's understanding, including that of a perceived paradox.

Paradox

Life's paradoxes are a "double vision" of sorts, like peering out of a house through a pane of window glass, thereby seeing objects that lie outside the house while simultaneously seeing reflections of things that lie within. Similarly, we can observe the workings of the outer world of Nature through physics and biology, simultaneously experiencing the inner workings and images of our own psychological maturity. This phenomenon is perhaps most clearly illustrated in the night sky, where stars and constellations bear names and images of our mythological heritage, concurrently serving as an entry into the scientific understanding of the biophysical universe.

Paradox comes from the Greek *parádoxos*, meaning "unbelievable." A paradox is represented by the sacred Hindu statue portraying the "Three Faces of God"—one oriented to the right, one oriented straight ahead, and one oriented to the left. The face oriented straight ahead is situated between, and joined to, those looking to the right and to the left. The pair of opposites (the face looking right and the one looking left) represents the apparent self-contradictory nature of a thing or situation—its appearance. The face sandwiched between the pair of opposites, on the other hand, represents the insight, the truth expressed by the union of the opposites.

What, you might ask, is the paradox of social-environmental sustainability? The paradox lies visibly in the apparent incongruity of Everycity as a human artifact *and* Everyforest as an ecosystem. What about their unity?

Unity

The unity of Everyforest and Everycity lies unseen in the commonality of their design. Put differently, to understand either, we must understand the design elements of both. To illustrate, let us examine two disparate concepts: (1) the similarities and differences of mosquitoes found worldwide and (2) the perceived, mutual exclusivity between "natural" and "artificial," in the sense of being made or altered by humans. First, let us discuss the mosquitoes.

If we compare ten individuals of the same kind of mosquitoes found in the city of Kuala Lumpur near the equator on the west coast of the Malaysian Peninsula with ten individuals of the same kind of mosquitoes found in the city of Cayenne, French Guiana, also near the equator but on the east coast of South America, we would probably classify the two groups as distinctly different species—a classical, clear-cut "either/or" categorization—based on their extreme divergence in appearance. But, if we were now to collect mosquitoes from around the world along the equator, what would we find when they were compared? That would depend on how we compared them.

If we compared each group of ten mosquitoes with their nearest neighbor (ten from location A with ten from location B), we would find an astonishing degree of similarity between the two samples. If we then compared ten from location B with ten from location C and ten from location C with ten from location D, we would find both continual similarities and gradual differences. Should we then compare samples of mosquitoes from location A with location D, we would begin to see increased divergence in characteristics.

So, instead of the neighboring samples of mosquitoes being unequivocally this or that because of clearly distinguishable characteristics, they form a continuum of gradually changing characteristics from A to Z. Despite how similar or dissimilar any two samples of mosquitoes may appear, the same examination will demonstrate the shared elements of design, such as the basic structure and function of their wings, antennae, mouth parts, body, and so on. When viewed as a continuum, the mosquitoes form a circle of similarities that reveal the commonalties of design elements necessary if they are all to be aerodynamically capable of flight and sucking blood, among other shared characteristics that make a mosquito a mosquito.

There is a social-environmental corollary to this phenomenon that most people fail to understand because they view the world as either "natural," meaning *undisturbed* by humans, or *artificial*, meaning either altered or made by humans. Much time is spent today arguing whether an ecosystem is ecologically natural or unnatural, economically good or bad. What if we thought of ecosystems or habitats as neither natural nor unnatural in the sense of either/or but rather as a continuum of naturalness?

For example, a mountaintop untouched by human alterations constitutes the most natural end of the continuum, whereas a shopping mall constitutes the most cultural end. Such a continuum can easily be symbolized as $N \leftrightarrow C$, where N represents the most natural end of the continuum and C the most cultural end. Everything in between, depending on where along the continuum it falls and how it is characterized, represents a degree of naturalness or a degree of culturalness.

The question for us today is: Where along this continuum must we, of necessity, maintain a piece of land in a given ecological condition if the whole of the landscape is to be sustainable, both environmentally and socially? And, it is here, concerning this question, the answer to which is determined by the collective of our individual choices, that people often disagree sharply with one another because each person has a different perspective as part of his or her frame of reference.

Everyone Is Right from His or Her Own Perspective

I now come to the notion of right versus wrong, again based on perceived similarities and differences. Society is composed of individual human beings much as the compound eye of an insect is composed of individual facets, each of which is slightly different in structure but equally important to the total vision of the eye. Each facet has its own light-sensitive element; each has its own refractive system, and each receives and transmits but a portion of the image.

As there are as many points of view in the compound eye of an insect as there are facets, so are there as many points of view in a society as there are people. Although everyone is right from his or her point of view, no one person has the complete image, and no one is totally correct.

We all sense differently the things we see, hear, touch, taste, and smell; because we sense things differently, we understand them differently. In addition, our senses are variously affected by the ever-changing circumstances of the world around us as well as the aging process to which we are each subjected. Our brains coordinate and integrate our ability to detect things and so produce our own sense of awareness. Through communication, our individual awareness is coordinated and integrated into the collective consciousness of many individuals. And, it is through our sensitivity that we become attuned to the complementary nature of one another's perceptions.

Differences between individual perceptions lead to the idea of right and wrong. While no one can convince you that you are wrong without somehow attacking your dignity, they can give you new data that may move you to make a different decision based on that information. In this way, you can be given the space that allows you to change your mind if you so choose while maintaining your dignity intact.

The question is: Who is right when we are all right from our own points of view? If everyone is right, then who is wrong? Because no one is wrong, we cannot argue any case—whether one is a product-oriented thinker or a systems-oriented thinker—based on "right" or "wrong." Right or wrong is always a human judgment dealing with appearances and a personal belief system, not reality (the level of water in a glass), which means that if one thinks he or she is right, that person must "win," and if the person wins, he or she is clearly right. The other person, therefore, is wrong because that person "lost" and lost because the person was surely wrong. With this kind of reasoning, each side becomes committed to winning agreement with its point of view and is not in a position to contemplate another possibility under the illusion of winners and losers.

But, the "virtual" duality of right versus wrong does not have to exist in reality. There is instead a continuum of "rightness" in which some are a little more right than others. This said, no one knows for sure who the "more correct" really is since no one person has the complete image, and thus no one person is totally right—although the person may think otherwise. Such a continuum is predicated on our individual lack of knowledge owing to our limited perception of possible outcomes. Since we cannot know for sure who is more right than whom, fairness necessitates that everyone is right from a personal perspective, and that each perspective is to some degree different—*not* wrong, regardless of the topic of discussion. Hence, the notion of "wrong" is unacceptable in resolving an issue.

If we are to survive the present upheavals of social evolution, we must be willing to accept the notion of *right, right,* and *different*. Wrong in the classical, combative sense must become a relic of the past if we are to treat others as we ourselves would like to be treated. Only if we can agree that everyone is right from a personal point of view will we be able to negotiate a common frame of reference. Only then can any issue be *resolved* in such a way that each side retains its dignity and society can progress with some semblance of order into the future. Only then will we be able to cooperate with one another in protecting the commonalties of design within and among ecosystems that are necessary to perpetuate our social-environmental sustainability.

Summation

In Chapter 1, I discussed the purely human dimension of our journey toward social-environmental sustainability: how we perceive the elements of design, how we share our perceptions, and the necessity of our interpersonal cooperation if society, as we know it, is to survive the duration of this century. Understanding how we communicate our sense of reality to one another is the topic of Chapter 2. Clear communication is critical if we are to design our villages, towns, and cities within the sustainable parameters of the varied landscapes that support them.

Notes

1. Chris Maser. *The Perpetual Consequences of Fear and Violence: Rethinking the Future.* Maisonneuve Press, Washington, DC (2004).
2. Mayo Clinic. Nicotine Addiction. http://www.americanheart.org/presenter.jhtml?identifier=4753 (accessed January 6, 2009); and Mayo Clinic. Nicotine Dependence. http://www.mayoclinic.com/health/nicotine-dependence/DS00307 (accessed January 6, 2009).
3. Wake Forest Univeristy Baptist Medical Center. What Drives Alcohol Addiction. http://www1.wfubmc.edu/articles/Alcohol+addiction (accessed January 6, 2009).
4. National Institute on Drug Abuse. Addiction: Drugs, Brains, and Behavior—The Science of Addiction. http://www.drugabuse.gov/Scienceofaddiction/ (accessed January 6, 2009); and Mayo Clinic. Drug Addiction. http://www.mayoclinic.com/health/drug-addiction/DS00183 (accessed January 6, 2009).
5. The National Council on Problem Gambling. http://www.ncpgambling.org/i4a/pages/index.cfm?pageid=1 (accessed January 6, 2009).
6. Junk Food, Addictions and the Obesity Problem. http://www.weightlossforall.com/junk-food-obesity-x.htm (accessed January 6, 2009).
7. Centers for Disease Control and Prevention. Overweight and Obesity. http://www.cdc.gov/nccdphp/dnpa/obesity/ (accessed January 6, 2009).
8. Garrett Hardin. Cultural carrying capacity: a biological approach to human problems. *BioScience*, 36 (1986):599–606.
9. The preceding two paragraphs on pharmaceutical wastes are based on D. G. Joakim Larsson, Cecilia de Pedro, and Nicklas Paxeus. Effluent from drug manufacturers contains extremely high levels of pharmaceuticals. *Journal of Hazardous Materials*, 148 (2007):751–755.
10. M. B. Almendro-Candel, M. M. Jordán, J. Navarro-Pedreño, and others. Environmental evaluation of sewage sludge application to reclaim limestone quarries wastes as soil amendments. *Soil Biology and Biochemistry*, 39 (2007):1328–1332.

11. Sources of Groundwater Contamination. http://www.groundwater.org/gi/sourcesofgwcontam.html (accessed January 6, 2009); Groundwater Contamination. http://www.epa.gov/superfund/students/wastsite/grndwatr.htm (accessed January 6, 2009); Preventing Groundwater Contamination. http://www.deq.state.mi.us/documents/deq-ead-tas-grwtrcon.pdf (accessed January 6, 2009).
12. Eugene P. Odum. *Ecology and Our Endangered Life Support Systems.* Sinauer Associates, Sunderland, MA (1989).
13. Albert Einstein. http://Science.prodos.org (accessed January 2, 2009).
14. Francis Bacon. http://Science.prodos.org (accessed January 2, 2009).
15. Sigmund Freud. *The Future of an Illusion.* W. W. Norton & Co., New York, (1961).
16. (no author given) Ancient Landmarks: The First Greek Philosophers. Theosophy 27 (1939):196–201. http://www.blavatsky.net/magazine/theosophy/ww/additional/ancientlandmarks/FirstGreekPhilosophers.html (accessed January 5, 2009).
17. Virginia M. Fellows. *The Shakespeare Code.* Summit University Press, Gardiner, MT (2006).
18. Fritjof Capra. *The Tao of Physics.* Shambhala, Berkeley, CA (1975).
19. William Blake Quotes. http://thinkexist.com/quotation/if_the_doors_of_perception_were_cleansed/179709.html (accessed January 15, 2009).
20. Louis Fischer. *Gandhi: His Life and Message for the World.* Penguin Group (USA) Inc., New York (1982).
21. Richard Attenborough. *The Words of Gandhi.* Newmarket Press, New York (1982).
22. Fischer. *Gandhi.*

2

Language and Communication

Introduction

Language is an important consideration because our conceptual design for social-environmental sustainability is formulated through language, the medium we use to communicate with one another as we commit our design to Everycity and so the collective landscape.

Every human language—the master tool representing a culture—has its unique construct that determines both its limitations and its possibilities in expressing myth, emotion, ideas, desires, and logic. The dreamers, artisans, and historians (the weavers of human mythology) use words to compose the broad shapes of a cultural story line. Language is the medium through which the spiritual condition of the human soul is painted and the material condition of human desire is wrought on the land. Words simultaneously set things in motion and influence their effects.

Words are also the basis of our reality by defining "what is." How we define something can have the effect of maintaining the status quo. To illustrate, some might say that the way our Western society defines *poverty* and *prosperity* perpetuates the social, economic, and environmental problems we are experiencing. If we continue to define *prosperity* in terms of a certain level of accumulated material wealth, we will continue to plunder natural resources because a higher "standard of living" (linked to prosperity) is something many, if not most, Americans seek. Hence, the lack of material wealth is defined as poverty when it falls at or below a certain level of acquisition.

But what if *poverty* was defined as the absence of friends, family, or spiritual wellness (instead of lacking the monetary means to acquire material possessions) and *prosperity* was defined as having an abundance of friends; a loving, functional family; and a fulfilling spiritual life (instead of a profusion of the monetary means whereby to acquire material possessions)? What would our society look like then? How about our economy?

The independence of cultural evolution creates the uniqueness cultures experience within themselves and within the reciprocity they experience with one another and with their immediate environments. Each culture, and each community within that culture, affects its environment in its own way and is accordingly affected by the environment in a particular way. So it is that people of different cultures each create a distinctive landscape that is, in some measure, reflected in the myths they hold and the languages they speak. As long as we have the maximum diversity of languages that cultural artists can use to paint verbal pictures of mythology, we can see ourselves—the collective human creature, the social animal—most clearly and from many points of view in a multitude of social mirrors.

The Process of Communication

Our human process of communication is composed of at least three elements: (1) the sender, someone speaking, writing, signing, exuding the silent language of attitude, or displaying art in its various forms; (2) the symbols used in creating and transmitting the message, sounds of a particular and repetitive form (spoken words); specific, repetitive hand-crafted signs (written words); a distinctive arrangement of musical notes (melody); facial expressions and "body language"; and (3) the receiver, someone listening to, reading, or observing the symbols. These elements are dynamically interrelated, and what affects one influences all through synergism.

This synergism makes communication a complicated, two-way process. If, for instance, a listener has difficulty understanding the symbols and indicates confusion, the speaker may become uncertain and timid and so lose confidence in being able to convey ideas. Hence, the effectiveness of the communication is diminished. On the other hand, when a listener reacts positively, a speaker is encouraged and adds strength and confidence to the message. Let us examine how the three elements work. For the sake of simplicity, only the spoken language (e.g., speaker and listener) is discussed.

Speaker

The originator of the message, or the speaker, probably plays the most critical role in the communication process because the speaker introduces something into the environment that cannot be withdrawn and is subject to an infinite range of interpretations. Thus, the speaker's awareness and understanding of the power of his or her internal language, as the basis for the message to be sent, is critical. Speech, or other forms of communication, is an outcome of an internal process that begins with a sensation or complex of sensations. Whether the expression that results is verbal or physical, both are actions.

First, we selectively attend to or sense something happening. Our body chemistry or the degree of wellness, our intention, and our overall state of mind influence this part of the process. Judgment or interpretation follows and results in perception or the definition of "what is." At this level, our internal language forms thoughts, and as we all know, for every thought, there is an effect. For instance, how might someone feel toward an acquaintance if he or she thought of the acquaintance as an "arrogant so and so"? Conversely, how might that same person feel if he or she wondered instead why the acquaintance was so "frightened and insecure"?

The latter thought is disarming, compared with the former, and will certainly result in a different form of expression in words or action. Thus, an awareness of our thoughts and the language we choose to express them can enhance our ability to communicate in a truly effective manner.

Listener

It is a basic rule of transmitting information that it is the responsibility of the speaker to be clear, concise, and relevant. But, since communication is a shared responsibility, listeners are obliged to do their best to assist in making the communication successful through understanding.

To understand the communication process, it helps to appreciate at least three aspects of listeners: their abilities, attitudes, and experiences. First, it is important to discern a listener's ability to question and comprehend the ideas transmitted, something that can be encouraged by providing a safe atmosphere that welcomes participation.

Second, a listener's attitude may be one of resistance, willingness, or passivity. Whatever the attitude, a speaker must gain the listener's attention and retain it. The more varied, interesting, and relevant a speaker is, the more successful the speaker will be in this respect.

Third, a listener's background, experience, and education constitute the frame of reference the speaker must understand. Here, it is critical for the speaker to assume the obligation of assessing the listener's knowledge and using it as the fundamental guide for effective communication. To get a listener's response, however, the speaker must first reach the listener, and it is here that major barriers to communication are usually found.

Barriers to Effective Communication

The nature of language and the way it is used often lead to misunderstandings and conflict that stem from four primary barriers to effective communication: (1) the lack of a common experience or frame of reference; (2) the attitude with which one approaches life (e.g., assertively or passively); (3) the use of abstractions; and (4) the inability to transfer experiences from one situation to another.

Lack of a Common Experience

The lack of a common experience or frame of reference, such as understanding the commonalties of design between an ecosystem and a city, is probably the greatest barrier to effective communication. Although many people believe that words carry meaning in much the same way as a person transports an armful of wood or a pail of water from one place to another, words *never* carry precisely the same meaning from the mind of the speaker to that of the listener. Words are vehicles of perceived meaning. They may or may not supply emotional meaning as well. The nature of the response is determined in large measure by the listener's past experiences surrounding each word and the feelings each word evokes.

These feelings grant a word meaning in the listener's mind but not in the word itself. Since a common frame of reference is basic to communication, words are meaningless in and of themselves. They are merely symbolic representations that correspond to anything to which people apply the symbol—objects, actions, experiences, or feelings.

Use of Abstractions

Concrete words refer to objects a person can directly experience, such as the cool of indoor air conditioning on a blistering hot day. Abstract words, on the other hand, represent ideas that lack exact boundaries of experience. Say a friend recounts the challenges of city planning above the Arctic Circle, where temperatures plunge to minus 50° Fahrenheit, something you had never encountered, how could you understand the image the friend conveyed? You could not because abstractions are shorthand symbols used to cover vast areas of experience or concepts that reach into the trackless, open-ended experiment of life. Albeit convenient and useful, abstractions can lead to misunderstandings.

The danger of using abstractions is their tendency to evoke an amorphous generality in the listener's mind and not the specific experience the speaker intended. In turn, the listener has no way of knowing what experiences the speaker wanted an abstraction to include.

When abstractions are used, they must be linked to specific experiences through examples, analogies, and illustrations. It is even better to use simple, concrete words with specific meanings. In this way, the speaker gains greater control of the images produced in the listener's mind, and language becomes a more effective tool. For instance, in comparing the functional design of Everyforest to that of Everycity, it is necessary to get the participants physically out of the comfortable conference room into both the forest and the city, where they can wander around and discuss what they perceive. This kind of experience helps transform the abstractions of the conference room into concrete examples in the real world, something everyone can see, touch, smell, hear, and, if necessary, taste. This kind of experience helps people to overcome their inability to transfer a principle from one situation to another.

Inability to Transfer Experiences from One Situation to Another

A major barrier to communication is the inability to transfer the outcomes of experience—in the form of a functional principle—from one kind of situation to another. The potential ability to transfer experiences from here to there is influenced by the relative breadth of one's experiences.

Experiential transfer is clearly critical to understanding how ecosystems and their interactive components function, including the bridge of reciprocity between a community and its surrounding environment. Analogies can be useful in aiding such transfer, as the following analogy demonstrates:

What happens when a process is "simplified"? The newly elected mayor of a city with an overspent budget guarantees to balance it. All that is necessary, the new mayor proposes, is eliminating services whose budgetary total accounts for the overexpenditure. What would happen, by way of example, if all police and fire services were eliminated? Would it make a difference, if the price were the same and the budget could still be balanced, if garbage collection was eliminated instead?

The problem with such a simplistic approach is in looking only at the cost of the terminated services and not at the functions they perform. To remove a piece of the whole may be acceptable, provided one knows which piece is being removed, what it does, and what effect—including cost over time—the loss of its function will have on the stability of the system. In an ecosystem, for instance, removing one piece, for which the long-term effect of its loss is unknown, may devastate the overall functionality of the system, which Aldo Leopold alluded to in 1933 when he wrote: "To build a better motor, we tap the uttermost powers of the human brain; to built a better countryside, we throw dice."[1]

Summation

Part I dealt with the purely human dimension of our social-environmental landscape and is not about the elements of design per se but rather about the way in which we *perceive* the elements and how we share our perceptions. Within this purview, Chapter 2 noted that effective communication is critical if we are to honor the parameters of Nature's biophysical principles and thus maintain our ecosystems intact while we plan our villages, towns, and cities—and so redesign our shared landscapes.

Part II is a brief treatise of Nature's rules of engagement in life and human behavior as they pertain to social-environmental planning. Growing out of the varied soils of culture, we are united by the hidden threads of our common human needs. If, therefore, we lose touch with one another as human beings, we will find a diminishing value in life. And, our common bonds will progressively erode into ever-increasing separateness. Unfortunately, technology, which originally made life easier and brought people together, is today increasingly disconnecting us (through mechanical distractions, myriad quick fixes, and symptomatic thinking) from the true nature of life and the life of Nature, which sustains us all. Here, Harrison E. Salisbury, American author and journalist, offered sagacious counsel: "There is no shortcut to life. To the end of our days, life is a lesson imperfectly learned."[2] To this, Marcus Aurelius (Roman Emperor from 161 to 180) would surely add, "Nothing has such power to broaden the mind as the ability to investigate systematically and truly all that comes under thy observation in life."[3]

To this end, Chapter 3 examines the inviolate biophysical principles that regulate how our world functions. Chapter 4, in turn, is a look at those principles of human behavior that allow a society to be functional, as opposed to dysfunctional, and thus foster a potential for people to live a life of dignity and well-being in greater harmony with their environment.

Notes

1. Aldo Leopold. *A Sand County Almanac*. Oxford University Press, New York (1966).
2. Harrison Evans Salisbury. Biography. http://www.answers.com/topic/salisbury-harrison (accessed January 3, 2009).
3. Marcus Aurelius. Inspiring Quotes from Great Scientists, Technologists, Inventors. http://science.prodos.org/quotes/index.html (accessed December 30, 2008).

Section II

Nature's Rules of Engagement in Social-Environmental Planning

3

The Law of Cosmic Unification

> In order to make an apple pie from scratch, you must first create the universe.
>
> **Astronomer and author Carl Sagan**[1]

Before time, the universe was naught, and, according to the Christian Bible, "The earth was without form, and void."[2] Then, according to current scientific thought, there arose a great cataclysm, the "big bang," and so was created a supremely harmonious and logical process as a foundation for the evolution of matter, from which the universe was born. So began the impartial process of evolution, a process that flows from the simple to the complex, from the general to the specific, and from the strongly bound to the weakly bound.

Although I suspect many people have at least some familiarity with the concept that evolution moves from the simple toward the complex and from the general toward the specific, I doubt as many people are familiar with the notion of moving from the strongly bound toward the weakly bound. To understand the last, envision a functional extended family. The strongest bond is between a husband and wife, then between the parents and their children. As the family grows, the bonds between the children and their various aunts and uncles and their first, second, and third cousins become progressively weaker as relationships become more distant with the increasing size of the family, not to mention the continual inclusion of marriage partners from heretofore unrelated families.

Taking this notion of the strength of a bond one step further, into a town or city, there is a definite limit to the number of people who can live together with a sense of community. This limit is brought about by the necessity of having frequent face-to-face contacts as a continuing bond of recognition. As a town, and particularly a large city, loses its unifying center, where people congregate, it commences to splinter into socially disjunct downtown areas and neighborhoods that often compete with one another for resources based on special interests.

Returning to the creation of the universe, it is necessary to examine its basic building blocks and the way they evolved into organized systems. The big bang created particles of an extremely high state of concentration that were bound together by almost unimaginably strong forces. From these original microunits, quarks and electrons were formed. (Scientists propose the term *quark* as the fundamental unit of matter.[3]) Quarks combined to form protons and neutrons; protons and neutrons formed atomic nuclei that were complemented by shells of electrons.[4] Atoms of various weights and complexities could, in some parts of the universe, combine into chains of molecules and, on suitable planetary surfaces, give birth to life. On Earth, for example, living organisms became ecological systems, wherein arose human communities with the remarkable features of language, consciousness, and a seemingly wide-ranging freedom of choice. Over time, these communities aggregated into societies with distinctive cultures.

In this giant process of evolution, relationships among things are in constant flux as complex systems arise from subatomic and atomic particles. In each higher level of complexity and organization, there is an increase in the size of the system and a corresponding decrease in the energies holding it together. Put differently, the forces that keep evolving systems intact, from a molecule to a human society, weaken as the size of the systems increases, yet the larger the system is, the more energy it requires in order to function. Such functional dynamics are characterized by their diversity as well as by the constraints of the overarching laws and subordinate principles that govern them.

I say these principles *govern* the world and our place in it because they form the behavioral constraints without which nothing could function in an orderly manner—especially social-environmental planning. In this sense, the Law of Cosmic Unification—the supreme law—is analogous to the Constitution of the United States, a central covenant that informs the subservient courts of each state about the acceptability of its governing laws. In turn, the Commons Usufruct Law represents the state's constitution, which instructs the citizens on acceptable behavior within the state. In this way, Nature's rules of engagement inform society of the latitude whereby it can interpret the biophysical principles and survive in a sustainable manner.

Understanding the Law of Cosmic Unification

The *Law of Cosmic Unification* is functionally derived from the synergistic effect of three universal laws: the first law of thermodynamics, the second law of thermodynamics, and the law of maximum entropy production.

The *first law of thermodynamics* states that the total amount of energy in the universe is constant, although it can be transformed from one form to another. Therefore, the amount of energy remains entirely the same, even if you could go forward or backward in time. For this reason, the contemporary notion of either "energy production" or "energy consumption" is a non sequitur. The *second law of thermodynamics* states that the amount of energy in forms available to do useful work can only diminish over time. The loss of available energy to perform certain tasks thus represents a diminishing capacity to maintain order at a certain level of manifestation (say, a tree) and so increases disorder or entropy. This "disorder" ultimately represents the continuum of change and novelty—the manifestation of a different, simpler configuration of order, such as the remaining ashes from the tree when it is burned. In turn, the *law of maximum entropy production* says that a system will select the path or assemblage of paths from available paths that minimizes the potential or maximizes the entropy at the fastest rate given the existing constraints.[5] The Law of Cosmic Unification is important to understand because social-environmental planning is about understanding and designing a flow chart of energy to accomplish a specific outcome.

The essence of maximum entropy simply means that, when any kind of constraint is removed, the flow of energy from a complex form to a simpler form speeds up to the maximum allowed by the relaxed constraint.[6] Clearly, we are all familiar with the fact that our body loses heat in cold weather, but our sense of heat lost increases exponentially when wind chill is factored into the equation because our clothing has ceased to be as effective a barrier to the cold—constraint to the loss of heat—it was before the wind became an issue. Moreover, the stronger and colder the wind, the faster our body loses its heat—the maximum entropy of our body's energy by which we stay warm. If the loss of body heat to the wind chill is not constrained, hypothermia and death ensue, along with the beginnings of bodily decomposition—reorganization from the complex structure and function toward a simpler structure and function.

In other words, systems are by nature dissipative structures that release energy by various means but inevitably by the quickest means possible. To illustrate, as a young forest grows old, it converts energy from the sun into living tissue that ultimately dies and accumulates as organic debris on the forest floor. There, through decomposition, the organic debris releases the energy stored in its dead tissue. Of course, rates of decomposition vary. A leaf rots quickly and releases its stored energy rapidly. Wood, on the other hand, generally rots more slowly, often over centuries in moist environments. As wood accumulates, so does energy stored in its fibers. Before the suppression of fires, they burned frequently enough to generally control the amount of energy stored in accumulating dead wood by burning it. These low-intensity fires protected a forest for decades, even centuries, from a catastrophic, killing fire. In this sense, a forest equates to a dissipative system in that energy acquired from the sun is released through the fastest means possible, be it gradually through decomposition or rapidly through a high-intensity fire. The ultimate constraint to the rate of entropic maximization, however, is the immediate weather in the short term and the overall climate in the long term.

Now, let us examine the notion of maximum entropy in a more familiar way. I have a wood-burning stove in my home with which I heat the 1,300 square feet of my living space. To keep my house at a certain temperature, I must control the amount of energy I extract from the wood I burn. I do this in nine ways.

My first consideration is the kind of wood I choose, be it Douglas-fir, western red cedar, western hemlock, bigleaf maple, Pacific madrone, Oregon ash, Oregon white oak, red alder, or a combination. My choice is important because each kind of wood has a different density and thus burns with a corresponding intensity. On one hand, the three coniferous woods (Douglas-fir, western red cedar, and western hemlock) are relatively soft, require less oxygen to burn than hardwoods, burn quickly, but produce only moderate heat. On the other hand, such hardwoods as bigleaf maple, Pacific madrone, Oregon ash, Oregon white oak, and red alder produce substantial heat—of which oak, madrone, and maple probably produce the most, followed by ash and alder. But, these hardwoods also require more oxygen to burn than the softwoods, and they burn more slowly.

The second concern is the quality of wood that I burn. Sound, well-seasoned wood burns far more efficiently than either wet, unseasoned wood or wood that is partially rotten. In this case, the quality of the wood also determines the effectiveness whereby it heats my house. Good-quality wood is far more effective in the production of heat than is wood of poor quality.

The third determination is the size and shape of the wood. Small pieces produce a lot of heat but are quick to disappear. Large pieces take more time to begin burning, but last longer and may or may not burn as hot when they really get going, depending on the kind of wood. Split wood has more surface area per volume and burns more rapidly than do round pieces of wood of the same size, such as large branches, because the latter have more volume than surface area.

The fourth decision is how wide to open the damper and thereby control the amount of air fanning the flames and therewith either increase the intensity of burning (opening the damper) or decrease the rate of burn (closing the damper). In each case, the length of time the damper is in a given position is part of the equation. The wider the damper is opened, the less the constraint, the hotter and faster the wood will burn, and the more rapidly heat will escape—the *law of maximum entropy production*. This law also addresses the speed with which wood is disorganized as wood and reorganized as ashes.

The fifth choice is how warm I want my house to be in terms of how cold it is outside. The colder it is outside, the more wood I must burn to maintain a certain level of heat—how much depends on the kind of wood I am burning. Conversely, the warmer it is outside, the less wood I must burn to maintain the same level of warmth.

The sixth consideration is how well my house is insulated against the intrusion of cold air and thus the escape of my indoor heat—both of which determine the amount of wood I must burn to maintain the temperature I want. Another facet of how much wood I must burn depends in part on whether clouds are holding the heat close to Earth, thus acting as a constraint to the heat leaking out of my house, or whether clear skies allow heat to bleed from my home and escape into outer space.

The seventh option is how often I open the outside door to go in and out of my house and so let cold air flow in to replace the warm air rushing out.

I could ameliorate this exchange by having an enclosed porch between the door opening into my house and the door opening directly to the outside. A well-insulated porch would act as a dead air space and would be a functional constraint to the loss of heat from my house as I access the outside.

The eight alternative is when to heat my house and for how long. I can, for instance, reduce the amount of heat I require at night when I am snuggled in bed. If I go to bed early and get up early, it is about the same as going to bed late and getting up late. But, if I go to bed early and get up late, I do not need to heat the house for as long as I would if I spent more time out of bed as opposed to in bed. Moreover, if it is well below freezing outside, I might have to keep the house warmer than otherwise to protect the water pipes from freezing.

And, the ninth course of action, one that is both influenced by the other eight and influences them in turn, is how warmly I choose to dress while indoors. Whatever I wear constitutes a constraint to heat loss of a greater or lesser degree. Clearly, the warmer I dress, the less wood I must burn in order to stay warm and vice versa. It is the same with how many blankets I have on my bed during the winter.

These nine seemingly independent courses of action coalesce into a synergistic suite of relationships wherein a change in one automatically influences the other eight facets of the speed wherewith energy from the burning wood escapes from my house. This said, the first and second laws of thermodynamics and the law of maximum entropy production meld to form the overall unifying law of the universe—the Law of Cosmic Unification—in which all subordinate principles, both biophysical and social, are encompassed. With respect to the functional melding of these three laws, Rod Swenson of the Center for the Ecological Study of Perception and Action, Department of Psychology, University of Connecticut, said that these three laws of thermodynamics "are special laws that sit above the other laws of physics as laws about laws or laws on which the other laws depend."[7] Stated a little differently, these three laws of physics coalesce to form the *supreme* Law of Cosmic Unification, to which all biophysical and social principles governing Nature and human behavior are subordinate—yet simultaneously *inviolate*. Inviolate means that we manipulate the effects of a principle through our actions on Earth, but we do not—and cannot—alter the principle itself.

The Inviolate Biophysical Principles

Although I have done my best to present the principles in a logical order, I find it difficult to be definitive because each principle forms an ever-interactive strand in the multidimensional web of energy interchange that constitutes the universe and our world within it. Moreover, I see a different possible order each time I read them, and each arrangement seems logical. Because each principle affects all principles, every arrangement is equally correct:

1. Everything is a relationship.
2. All relationships are productive.
3. The only true investment is energy from sunlight.
4. All systems are defined by their function.
5. All relationships result in a transfer of energy.
6. All relationships are self-reinforcing feedback loops.
7. All relationships have one or more trade-offs.
8. Change is a process of eternal becoming.
9. All relationships are irreversible.
10. All systems are based on composition, structure, and function.
11. All systems have cumulative effects, lag periods, and thresholds.
12. All systems are cyclical, but none is a perfect circle.
13. Systemic change is based on self-organized criticality.
14. Dynamic disequilibrium rules all systems.

Principle 1: Everything Is a Relationship

All we humans do—ever—is practice relationships because the existence of everything in the universe is an expression of its relationship to everything else. Moreover, all relationships are forever dynamic and thus constantly changing, from the wear on your toothbrush from daily use to the rotting lettuce you forgot in your refrigerator. Herein lies one of the foremost paradoxes of life: The ongoing process of change is a universal constant over which, much to our dismay, we have no control.

Think, for example, what the difference is between a motion picture and a snapshot. Although a motion picture is composed of individual frames (instantaneous snapshots of the present moment), each frame is entrained in the continuum of time and thus cannot be held constant, as Roman Emperor Marcus Aurelius observed: "Time is a river of passing events, and strong is its current. No sooner is a thing brought to sight than it is swept by and another takes its place, and this too will be swept away."[8]

Yet we, in our fear of uncertainty, are continually trying to hold the circumstances of our life in the arena of constancy as depicted in a snapshot—hence, the frequently used term *preservation* in regard to this or that ecosystem, this or that building. Yet, jams and jellies are correctly referred to as "preserves" because they are heated during their preparation to kill all living organisms and thereby prevent noticeable change in their consistency.

Insects in amber are an example of true preservation in Nature. Amberization, the process whereby fresh resin is transformed into amber, is so gentle that it forms the most complete type of fossilization known for small, delicate, soft-bodied organisms, such as insects. In fact, a small piece of amber found along the south coast of England in 2006 contained a 140-million-year-old spider web constructed in the same orb configuration as that of today's garden spiders. This is 30 million years older than a previous spider web found encased in Spanish amber. The web demonstrates that spiders have been ensnaring their prey since the time of the dinosaurs. And, because amber is three dimensional in form, it preserves color patterns and minute details of the organism's exoskeleton and so allows the study of microevolution, biogeography, mimicry, behavior, reconstruction of the environmental characteristics, the chronology of extinctions, paleosymbiosis,[9] and molecular phylogeny.[10] But, the same dynamic cannot be employed outside an airtight container, such as a drop of amber or canning jar. In other words, whether natural or artificial, all functional systems are open because they all require the input of a sustainable supply of energy in order to function; conversely, a totally closed, functional system is a physical impossibility.

Principle 2: All Relationships Are Productive

I have often heard people say that a particular piece of land is "unproductive" and needs to be "brought under management." Here, it must be rendered clear that every relationship is productive of a cause that has an *effect*, and the effect, which is the cause of another effect, *is* the product. Therefore, the notion of an unproductive parcel of ground or an unproductive political meeting is an illustration of the narrowness of human valuation because such judgment is viewed strictly within the extrinsic realm of personal values, usually economics—not the intrinsic realm of Nature's dynamics that not only transcend our human understanding but also defy the validity of our economic assessments.

We are not, after all, so powerful a natural force that we can destroy an ecosystem because it still obeys the biophysical principles that determine how it functions at any point in time. Nevertheless, we can so severely alter an ecosystem that it is incapable of providing—for all time—those goods and services we require for a sustainable life. Bear in mind that the total surface area of the United States covered in paved roads precludes the soil's ability to capture and store water or that we are currently impairing the ocean's ability to sequester carbon dioxide (one of the main greenhouse gases) because we have so dramatically disrupted the population dynamics of the marine fishes by systematically overexploiting too many of the top predators.[11] All of the relationships that we affect are productive of some kind of outcome—a product. Now, whether the product is beneficial for our use or even amenable to our existence is another issue.

Principle 3: The Only True Investment Is Energy from Sunlight

The only true investment in the global ecosystem is energy from solar radiation (materialized sunlight); everything else is merely the recycling of already existing energy. In a business sense, for example, one makes money (economic capital) and then takes a percentage of those earnings and *recycles* them, puts them back as a cost into the maintenance of buildings and equipment in order to continue making a profit by protecting the integrity of the initial outlay of capital over time. In a business, one recycles economic capital *after* the profits have been earned.

Biological capital, on the other hand, must be "recycled" *before* the profits are earned. This means forgoing some potential monetary gain by leaving enough of the ecosystem intact for it to function in a sustainable manner. In a forest, for instance, one leaves some proportion of the merchantable trees (both alive and dead) to rot and recycle into the soil and thereby replenish the fabric of the living system. In rangelands, one leaves the forage plants in a viable condition so they can seed and protect the soil from erosion as well as add organic material to the soil's long-term, ecological integrity.

People speak incorrectly about fertilization as an *investment* in a forest or grassland, when in fact it is merely recycling chemical compounds that already exist on Earth. In reality, people are simply taking energy (in the form of chemical compounds) from one place and putting them in another for a specific purpose. The so-called investments in the stock market are a similar shuffling of energy.

When people *invest* money in the stock market, they are really recycling energy from Nature's products and services that were acquired through human labor. The value of the labor is transferred symbolically to a dollar amount, thereby representing a predetermined amount of labor. Let us say you work for ten dollars an hour; then, a one-hundred-dollar bill would equal ten hours of labor. Where is the *investment*? There is no investment, but there is a symbolic recycling of the energy put forth by the denomination of money we spend.

Here, you might argue that people *invested* their labor in earning the money. And, I would counter that whatever energy they put forth was merely a recycling of the energy they took in through the food they ate. Nevertheless, the energy embodied in the food may actually have simultaneously been a true investment and a recycling of already existing energy.

It has long been understood that green plants use the chlorophyll molecule to absorb sunlight and use its energy to synthesize carbohydrates (in this case, sugars) from carbon dioxide and water. This process is known as photosynthesis, where *photo* means "light" and *synthesis* means the "fusion of energy" and is the basis for sustaining the life processes of all plants. The energy is derived from the sun (an original input) and combined with carbon dioxide and water (existing chemical compounds) to create a renewable source of usable energy. This process is analogous to an array of organic solar panels—the green plant.

Think of it this way, the plant (an array of solar panels) uses the green chlorophyll molecule (a *photoreceptor,* meaning receiver of light) to collect light from the sun within chloroplasts (small, enclosed structures in the plant that are analogous to individual solar panels). Then, through the process of photosynthesis, the sun's light is used to convert carbon dioxide and water to carbohydrates for use by the plant, a process that is comparable to converting the sun's light in solar panels on the roof of a building into electricity for our use. These carbohydrates, in turn, are partly stored energy from the sun—a new input of energy into the global ecosystem—and partly the storage of existing energy from the amalgam of carbon dioxide and water.[12]

When, therefore, we eat green plants, the carbohydrates are converted through our bodily functions into different sorts of energy. By that I mean the energy embodied in green plants is altered through digestion into the various types of energy our bodies require for their physiological functions. The *excess* energy (that not required for physiological functions) is expended in the form of physical motion, such as energy to do work. On the other hand, it is different when eating meat because the animal has already used the sun's contribution to the energy matrix in its own bodily functions and its own physical acts of living, so all we get from eating flesh is recycled energy.

Principle 4: All Systems Are Defined by Their Function

The behavior of a system—any system—depends on how its individual parts interact as functional components of the whole, not on what an isolated part is doing. The whole, in turn, can only be understood through the relationships, the interaction of its parts. The only way anything can exist is encompassed in its interdependent relationship to everything else, a physical limitation that means an isolated fragment or an independent variable can exist *only on paper* as a figment of the human imagination.

Put differently, the false assumption is that an independent variable of one's choosing can exist in a system of one's choice, and that it will indeed act as an independent variable. In reality, all systems are interdependent and thus rely on their pieces to act in concert as a functioning whole. This being the case, no individual piece can stand on its own *and* simultaneously be part of an interactive system. Thus, *there neither is nor can there be an independent variable* in any system, be it biological, biophysical, or mechanical because every system is interactive by its very definition as a system.

What is more, every relationship is constantly adjusting itself to fit precisely into other relationships that, in turn, are consequently adjusting themselves to fit precisely into all relationships, a dynamic that *precludes* the existence of a constant value of anything at anytime. Hence, to understand a system as a functional whole, we need to understand how it fits into the larger system of which it is a part and so gives us a view of systems supporting systems supporting systems supporting systems, ad infinitum.

Principle 5: All Relationships Result in a Transfer of Energy

Although technically a "conduit" is a hollow tube of some sort, I use the term here to connote any system employed specifically for the transfer of energy from one place to another. Every living thing, from a virus to a bacterium, fungus, plant, insect, fish, amphibian, reptile, bird, and mammal, is a conduit for the collection, transformation, absorption, storage, transfer, and expulsion of energy. In fact, the function of the entire biophysical system is tied up in the collection, transformation, absorption, storage, transfer, and expulsion of energy—one gigantic, energy-balancing act.

Principle 6: All Relationships Are Self-Reinforcing Feedback Loops

Everything in the universe is connected to everything else in a cosmic web of interactive feedback loops, all entrained in self-reinforcing relationships that continually create novel, never-ending stories of cause and effect, stories that began with the eternal mystery of the original story, the original cause. Everything, from a microbe to a galaxy, is defined by its ever-shifting relationship to every other component of the cosmos. Thus, "freedom" (perceived as the lack of constraints) is merely a continuum of fluid relativity. In contraposition, every relationship is the embodiment of interactive constraints to the flow of energy—the very dynamic that perpetuates the relativity of freedom and thus of all relationships.

Hence, every change (no matter how minute or how grand) constitutes a systemic modification that produces novel outcomes. A feedback loop, in this sense, comprises a reciprocal relationship among countless bursts of energy moving through specific strands in the cosmic web that cause forever-new, compounding changes at either end of the strand, as well as every connecting strand.[13] Here, we often face a dichotomy with respect to our human interests.

On the one hand, while all feedback loops are self-reinforcing, their effects in Nature are neutral because Nature is impartial with respect to consequences. We, on the other hand, have definite desires involving outcomes and thus assign a preconceived value to what we think of as the end result of Nature's biophysical feedback loops. A simple example might be the response of North American elk in the Pacific Northwestern United States to the alteration of their habitat. In this case, the competing values were (and still are) elk as an economically important game animal versus timber as an economically important commodity.

When I was a boy in the 1940s and 1950s, the timber industry coined the adage: Good timber management is good wildlife management. At the time, that claim seemed plausible because elk populations were growing in response to forests being clear-cut. By the mid- to late-1960s and throughout the 1970s, however, elk populations began to exhibit significant declines. Although predation was run out as the obvious reason, it did not hold up under scrutiny since the large predators, such as wolves and grizzly bears, had long been extirpated, and the mountain lion population had been decimated because of the bounties placed on the big cats.

As it turned out, the cause of the decline in elk numbers was subtler and far more complicated than originally thought. The drop in elk numbers was in direct response to habitat alteration by the timber industry. This is not surprising since elk, like all wildlife, have specific habitat requirements that consist of food, water, shelter, space, privacy, and the overall connectivity of the habitat that constitutes these features. When any one of these elements is in short supply, it acts as a limiting factor or constraint with respect to the viability of a species' population as a whole.

By way of illustration, here is a simplified example. In the early days before extensive logging began, the land was well clothed in trees, making food the factor that limited the number of elk in an area. As logging cleared large areas of forest, grasses and forbs grew abundantly; elk, being primarily grazers, became increasingly numerous. This relationship continued for some years, until—for an instant in time—the perfect balance between the requirements of food and shelter was reached. The proximity to water did not play as important a role in this balance because of the relative abundance of forest streams and because elk can travel vast distances to find water. Thus, hunters and loggers initially perceived clear-cut logging as the proverbial win-win situation (a positive, self-reinforcing feedback loop).

But, as it turned out, the main interplay among the potential limiting factors for elk was between food and shelter. At first, food was the limiting factor because elk were constrained in finding their preferred forage by the vast acres of contiguous forest. In contraposition, continued logging started to shift the habitat configuration in a way that proved detrimental to the elk because, while the habitat for feeding continued to increase with clear-cutting, that for shelter declined disproportionately. Accordingly, the shelter once provided by the forest became the factor that increasingly reversed the elk's growth in numbers. Here, it must be understood that shelter for elk consists of two categories—one for hiding in the face of potential danger (simply called *hiding cover*) and one for regulating the animal's body temperature (called *thermal cover).*

Thermal cover often consists of a combination of forest thickets or stands of old trees coupled with topographical features that block the flow of air. As such, thermal cover allows the elk to cool their bodies in dense shade in summer and get into areas of calm, out of the bitter winds, in winter, which markedly reduces the wind chill factor and thus conserves their body heat.[14] At length, the hunters began to see the systematic, widespread clear-cutting of the forest as a losing situation for huntable populations of elk (a negative, self-reinforcing feedback loop), although they did not equate the loss of thermal cover as the cause.

Another example of a self-reinforcing feedback loop is offered by the Dusky farmerfish around the Japanese islands of Ryukyu, Sesoko, and Okinawa. Dusky farmerfish establish and maintain monocultural farms of the red algae (seaweed known as filamentous rhodophytes) by defending them against invading grazers and by weeding out indigestible algae. To control their monocultures, the fish bite off the undesirable species of algae, swim to the edge of their territorial farms, and spit out the unwanted "weeds." Because the crops of red algae grow only in fish-tended monocultures, they die out if a farmerfish is removed from its farm. This in turn makes the algae's survival dependent on the ability of a fish to maintain its farm. Since this is the only algae harvested and eaten by the fish as its staple food, the reciprocal feedback loop is one of obligatory cultivation for mutual benefit.[15] In addition to simply maintaining a monocultural algae farm, however, the farmerfish inadvertently create a distinctive habitat that maintains and enhances a multispecies coexistence of foraminifera.[16]

These samples of feedback loops, like all others, are ultimately controlled by Earth's climate and so greatly influenced by the levels of atmospheric carbon dioxide (CO_2) over time. Evidence from ice cores and marine sediments indicate that changes in carbon dioxide over timescales beyond the glacial cycles are finely balanced and act to stabilize global temperatures.[17] What is more, the long-term balance between the emissions of carbon dioxide into the atmosphere through such events as volcanic eruptions and the removal of carbon dioxide from the atmosphere through such processes as its burial in deep-sea sediments holds true despite glacial–interglacial variations on relatively short timescales. Today, on the other hand, that part of the feedback loop whereby carbon dioxide is removed from the atmosphere by the chemical breakdown of silicate rock in mountains (termed *weathering*), as well as carbonate minerals (those containing CO_3) that are buried in deep-sea sediments, is being severely disrupted—even overwhelmed—by human activities that are raising the level of CO_2 emissions.[18]

Principle 7: All Relationships Have One or More Trade-Offs

As with the elk and farmerfish, all relationships have a trade-off that may be neither readily apparent nor immediately understood. To illustrate, for most of the past nine hundred years, the buildings in London were clean, many with cream-colored limestone façades. But, then things began to change as a result of the introduction of coal-burning stoves. That notwithstanding, the rate of change was so slow the cumulative effects were not readily apparent until a threshold of visibility had been crossed, and the protracted exposure to the sooty pollution of city air began to turn the buildings dark gray and black. And so it is that smutty buildings dominated the cities of Europe and the United States for most of the nineteenth and twentieth centuries.[19] In fact, archival photographs show that the limestone Cathedral of Learning on the

University of Pittsburgh campus in Pennsylvania, built during a period of heavy pollution in the 1930s, became soiled while still under construction.[20]

Reductions in Pittsburgh's air pollution began in the late 1940s and 1950s.[21] Since then, rain has slowly washed the soiled areas of the forty-two-story Cathedral of Learning, leaving a white, eroded surface. The patterns of whitened areas in archival photographs show the greatest rates of cleansing occurred on the corners of the high elevations on the building, predominantly where the impact of both rain and wind is most intense. It is also clear that the discoloration of buildings is a dynamic process by which the deposition of pollution is a relatively consistent process but is simultaneously washed away to varying degrees and patterns over the building's surface. Moreover, sooty pollutants soiled buildings, such as the Cathedral of Learning, much more rapidly in the past than they are being cleaned by wind and rain in the present.[22]

In this century, though, the buildings will gradually become more colorful as the city air is cleaned through the promulgation of pollution-control laws and windswept rain that will wash away the encrusted soot. The outcome of such cleaning may well be multicolored buildings as the natural reddish of some limestone is accentuated or a yellowing process that occurs as a result of pollutants that are more organic in constitution. What is more, the switch from coal to other fuels has cast the Tower of London in hues that are slightly yellow and reddish-brown. As the atmosphere is cleaned and thus dominated more by organic pollutants, a process of yellowing on stone buildings due to the oxidation of organic compounds in the fumes of diesel and gasoline may become of concern.[23] The oxidation of this increased organic content from the exhaust of motor vehicles may have overall aesthetic consequences for the management of historic buildings—namely, recognizing a shift away from the simple gypsum crusts of the past to those richer in organic materials and thus warmer tones, particularly browns and yellows.[24]

And, this says nothing about plant life growing on cleansed buildings, a phenomenon made possible because vehicular exhaust emits less of the sulfates that are present in the pollution from coal, pollution that suppress the growth of algae, lichens, and mosses. Consequently, buildings may come to exhibit greens, yellows, and reddish-brown in different places and various patterns because, while lichens and algae prefer humid environs, such as cracks, they can grow on flat surfaces as well.[25]

The foregoing deals only with the dynamics of Nature in response to soiling such limestone buildings as the Cathedral of Learning by different types of pollution and the long-term cleansing effects of wind and rain. Added to the trade-offs among these variables is the diversity of preferences espoused in 2003 by employees of the university.

Whereas some university officials were in favor of scrubbing the building with baking soda to remove the black, 70-year-old industrial grime, Cliff Davidson, the environmental engineer from Carnegie Mellon University who studied the building, prefers to let Nature do the work. Although, according to Davidson, the whiter spots have been scrubbed by wind-driven rain over decades, the darker spots in nooks and crannies might well remain for centuries, if they could be cleaned at all. In contrast, Doris Dyen, director of cultural conservation for the Rivers of Steel National Heritage Area, expressed appreciation of how buildings in Pittsburgh were being spruced up. "At the same time," she said, "you can lose a little bit of a sense of what Pittsburgh was like for 100 years when all the buildings were showing the effects of the 24-hour-a-day operation of the steel mills in the area." G. Alec Stewart, dean of the University Honors College, took yet a different tack: "It would make a stunning addition to the night skyline of Pittsburgh if we were able to illuminate it [the Cathedral of Learning] as significant monuments are in other major cities," comparing it to the Washington Monument.[26]

So, what are some of the significant trade-offs with respect to the Cathedral of Learning?

1. Clean the building artificially in the short term *or* let Nature do it over time.
2. Clean the building to blend into the cityscape and thus forgo the sense of familiarity *or* maintain the soot-derived appearance and thus avoid rapid change.
3. Trade the sooty, vegetation-free exterior for an exhaust-enriched, vegetation-covered exterior of the building and thereby give up a sense of Pittsburgh's one-hundred-year history.
4. Illuminate the Cathedral of Learning from the outside to create a monument-like effect, such as the Washington Monument in the District of Columbia *or* keep the status quo.

In the end, each of these trade-offs is couched in terms of whether to change or not, based largely on some cultural value that blends naturally into an emotional criterion.

Other relationships have much more discernable trade-offs. Take the springtime ozone hole over Antarctic as illustrative; it is finally shrinking after years of growing. As the hole grew in size due to the human-induced, ozone-destroying chemicals in the stratosphere, the risk of skin cancer increased because more ultraviolet radiation reached Earth. Although today the good news is that the ozone hole is now shrinking and, through a complicated cascade of effects, could fully close within this century, what about tomorrow? Because the hole in the stratospheric ozone layer does not absorb much ultraviolet radiation, it keeps the temperature of Antarctica much cooler than normal. A completely recovered ozone layer, on the other hand, could significantly boost atmospheric warming over and around the icy continent and ostensibly augment its melting.[27] In this case, what is good for humans may not be good for Antarctica and vice versa.

Principle 8: Change Is a Process of Eternal Becoming

Change, as a universal constant, is a continual process of inexorable novelty. It is a condition along a continuum that may reach a momentary pinnacle of harmony within our senses. Then, the very process that created the harmony takes it away and replaces it with something else—always with something else. Change requires constancy as its foil in order to exist as a dynamic process of eternal becoming. Without constancy, change could neither exist nor be recognized.

We all cause change of some kind every day. I remember a rather dramatic one I inadvertently made along a small stream flowing across the beach on its way to the sea. The stream, having eroded its way into the sand, created a small undercut that could not be seen from the top. Something captured my attention in the middle of the stream, and I stepped on the overhang to get a better look, causing the bank to cave in and me to get a really close-up view of the water. As a consequence of my misstep, I had both altered the configuration of the bank and caused innumerable grains of sand to be washed back into the sea from whence they had come several years earlier riding the crest of a storm wave.

Whereas mine was a small, personally created change in an infinitesimal part of the world, others are of gigantic proportions in their effects. People of civilizations that collapsed centuries ago are a good example of such gargantuan effects because they were probably oblivious to the impact that could be wrought by long-term shifts in climate. Although not likely to end the debate regarding what caused the demise of the Roman and Byzantine empires, new data suggest that a shift in climate may have been partly responsible. The plausibility of this notion has been given a scientific boost of credibility through studying the stalactites of Soreq Cave in Israel.[28]

Stalactites are the most familiar, bumpy, relatively icicle-shaped structures found hanging from the ceilings of limestone caves. They are formed when water accumulates minerals as it percolates through soil before seeping into a cave. If the water's journey takes it through limestone, it typically leaches calcium carbonate and carbon dioxide in its descent. The instant the water seeps from the ceiling of a cave, some of the dissolved carbon dioxide in the fluid escapes into the cave's air. This gentle, soda-pop-like fizzing process causes the droplet to become more acidic and so results in some of the calcium carbonate crystallizing on the cave's ceiling, thereby initiating a stalactite. As this process is performed over and over, the separation of calcium carbonate from within the thin film of fluid flowing down its surface allows the stalactite to grow. The procedure is so slow it typically takes a century to add four-tenths of an inch (one centimeter) to a stalactite's growth.[29] Moreover, stalactites, like tree rings, can tell stories of paleoclimatic events, such as the severe drought that took place on the Colorado Plateau in the mid-1100s.[30]

By using an ion microprobe, it has become possible to read the chemical deposition rings of the Soreq Cave stalactites with such precision that even seasonal increments of growth can be teased out of a given annual ring. The results indicate that a prolonged drought, beginning in the Levant region as far back as 200 years BCE and continuing to AD 1100, coincides with the fall of both Roman and Byzantine empires. (Levant is the former name of that region of the eastern Mediterranean that encompasses modern-day Lebanon, Israel, and parts of Syria and Turkey.) Although determining why civilizations collapse is always more complicated than one might imagine, an inhospitable shift in climate might well be part of the equation that either forces people to adapt by changing their behavior or eliminates them.[31] The latter seems to be the case in China.

The historical record of the Asian monsoon's activity is archived in an 1,800-year-old stalagmite found in Wanxiang Cave in the Gansu Province of north-central China. Mineral-rich waters dripping from the cave's ceiling onto its floor year after year formed the stalagmite (a mirror image of a stalactite) that grew continuously for 1,800 years, from AD 190 to 2003. Like trees and the stalactites in the Soreq Cave of Israel, stalagmites have annual growth rings that can provide clues about local environmental conditions for a particular year. Chapters in the Wanxiang Cave stalagmite, written over the centuries, tell of variations in climate that are similar to those of the Little Ice Age, Medieval Warm Period, and the Dark Age Cold Period recorded in Europe. Warmer years were associated with stronger East Asian monsoons.

By measuring the amount of oxygen-18 (a rare form of "heavy" oxygen) in the stalagmite's growth rings, the years of weak summer monsoons with less rain can be pinpointed due to the large amounts of oxygen-18 in the rings. The information secreted within the life of the stalagmite tells the story of strong and weak monsoons, which in turn chronicle the rise and fall of several Chinese dynasties. This is an important deliberation because monsoon winds have for centuries carried rain-laden clouds northward from the Indian Ocean every summer, thereby providing nearly 80 percent of the annual precipitation between May and September in some parts of China—precipitation critical to the irrigation of crops.

In periods when the monsoons were strong, dynasties, such as the Tang (AD 618–907) and the Northern Song (AD 960–1127), enjoyed increased yields of rice. In fact, the yield of rice during the first several decades of the Northern Song dynasty allowed the population to increase from 60 million to as many as 120 million. But, periods of weak monsoons ultimately spelled the demise of dynasties.

The Tang dynasty, for example, was established in AD 618 and is still determined to be a pinnacle of Chinese civilization, a kind of golden age from its inception until the ninth century, when the dynasty began to lose its grip. The Tang was dealt a deathblow in AD 873, when a growing drought turned horrific, and widespread famine took a heavy toll on both people and livestock. Henceforth, until its demise in AD 907, the Tang dynasty was plagued by civil unrest.

Weak monsoon seasons, when rains from the Indian Ocean no longer reached much of central and northern China, coincided with droughts and the declines of the Tang, Yuan (AD 1271–1368), and Ming (AD 1368–1644) dynasties, the last two characterized by continual popular unrest. Weak monsoons with dramatically diminished rainfall may also have helped trigger one of the most tumultuous eras in Chinese history, called the Five Dynasties and Ten Kingdoms period, during which time, five dynasties rose and fell within a few decades, and China fractured into several independent nation-states.

Data from the stalagmite indicate that the strength of past Asian monsoons was driven by the variability of natural influences—such as changes in solar cycles and global temperatures—until 1960, when anthropogenic activity appears to have superseded natural phenomena as the major driver of the monsoon seasons from the late twentieth century onward. In short, the Asian-monsoon cycle has been disrupted by human-caused climate change.[32] Here, an observation by the British biologist Charles Darwin is apropos: "It is not the strongest of the species that survive, nor the most intelligent, but the one most responsive to change."[33]

Principle 9: All Relationships Are Irreversible

Because change is a constant process orchestrated along the interactive web of universal relationships, it produces infinite novelty that precludes anything in the cosmos from ever being reversible. Take my misstep on the aforementioned stream's edge. One moment I was standing on the level beach, and the next I was conversing with the water. At the same time, the sand I had knocked into the stream was being summarily carried off to the sea. What of this dynamic was reversible? Nothing was reversible because I could not go back in time and make a different decision of where to place my foot. And, because we cannot go back in time, nothing can be restored to its former condition. All we can ever do is repair something that is broken so it can continue to function, albeit differently from its original form. If you want a detailed discussion of this principle, read *Earth in Our Care*.[34]

Principle 10: All Systems Are Based on Composition, Structure, and Function

We perceive objects by means of their obvious composition, structure, or function. *Structure* is the configuration of elements, parts, or constituents of something, be it simple or complex. The structure can be thought of as the organization, arrangement, or make-up of a thing. *Function*, on the other hand, is what a particular structure either can do or allows to be done to it or with it.

Let us examine a common object, a chair. A chair is a chair because its structure gives it a particular shape. A chair can be characterized as a piece of furniture consisting of a seat, four legs, and a back; it is an object designed to accommodate a sitting person. If we add two arms, we have an *armchair* wherein we can sit and rest our arms. Should we now decide to add two rockers to the bottom of the chair's legs, we have a *rocking chair* in which we can sit, rest our arms, and rock back and forth while doing so. Nevertheless, it is the seat that allows us to sit in the chair, and it is the act of sitting, the functional component allowed by the structure, that makes a chair, a chair.

Suppose we remove the seat so the structure that supports our sitting no longer exists. Now to sit, we must sit on the ground between the legs of the once-chair. By definition, when we remove a chair's seat, we no longer have a chair because we have altered the structure and therefore altered its function. Thus, the structure of an object defines its function, and the function of an object defines its necessary structure. How might the interrelationship of structure and function work in Nature?

To maintain ecological functions means that one must maintain the characteristics of the ecosystem in such a way that its processes are sustainable. The characteristics one must be concerned with are (1) composition, (2) structure, (3) function, and (4) Nature's disturbance regimes that periodically alter an ecosystem's composition, structure, and function.

We can, for example, change the composition of an ecosystem, such as the kinds and arrangement of plants in a forest or grassland; this alteration means that composition is malleable to human desire and thus negotiable within the context of cause and effect. In this case, composition is the determiner of the structure and function in that composition is the cause, rather than the effect, of the structure and function.

Composition determines the structure, and structure determines the function. Thus, by negotiating the composition, we simultaneously negotiate both the structure and function. On the other hand, once the composition is in place, the structure and function are set—unless, of course, the composition is altered, at which time both the structure and function are altered accordingly.

Returning momentarily to the chair analogy, suppose you have an armchair in which you can sit comfortably. What would happen if you either gained a lot of weight or lost a lot of weight but the size of the chair remained the same? If, on the one hand, you gained a lot of weight, you might no longer fit into your chair. On the other hand, if you lost much weight, the chair might be uncomfortably large. In the first case, you could alter the composition by removing the arms and thus be able to sit on the chair. In the second case, you might dismantle the chair, replace the large seat with a smaller one, and reassemble the chair.

In a similar but more complex fashion, the composition or kinds of plants and their age classes within a plant community create a certain structure that is characteristic of the plant community at any given age. It is the structure of the plant community that in turn creates and maintains certain functions. In addition, it is the composition, structure, and function of a plant community that determine what kinds animals can live there, how many, and for how long.

Hence, if one changes the composition of a forest, one changes the structure, hence the function, and thus affects the animals. The animals in general are not just a reflection of the composition but ultimately constrained by it.

If townspeople want a particular animal or group of animals within its urban growth boundary, let us say a rich diversity of summering birds and colorful butterflies to attract tourist dollars from bird-watchers and tourists in general, members of the community would have to work backward by determining what kind of function to create. To do so, they would have to know what kind of structure to create, which means knowing what type of composition is necessary to produce the required habitat for the animal the community wants. Thus, once the composition is ensconced, the structure and its attendant functions operate as an interactive unit in terms of the habitat required for the animal.

People and Nature are continually changing the structure and function of this ecosystem or that ecosystem by manipulating the composition of its plants, an act that subsequently changes the composition of the animals dependent on the structure and function of the resultant habitat. By altering the composition of plants within an ecosystem, people and Nature alter its structure and in turn affect how it functions, which in turn determines not only what kinds of individuals and how many can live there but also what uses humans can make of the ecosystem.

Principle 11: All Systems Have Cumulative Effects, Lag Periods, and Thresholds

Nature, as I have said, has intrinsic value only and so allows each component of an ecosystem to develop its prescribed structure, carry out its ecological function, and interact with other components through their evolved, interdependent processes and self-reinforcing feedback loops. No component is more or less important than another; each may differ from the other in form, but all are complementary in function, which operates through cumulative effects, lag periods, and thresholds.

Our intellectual challenge is recognizing that no given factor can be singled out as the sole cause of anything. All things operate synergistically as cumulative effects that exhibit a lag period before fully manifesting themselves. Cumulative effects, which encompass many little, inherent novelties, cannot be understood statistically because ecological relationships are far more complex and far less predictable than our statistical models lead us to believe—a circumstance that Francis Bacon may have been alluding to when he said, "The subtlety of Nature is greater many times over than the subtlety of the senses and understanding."[35] In essence, Bacon's observation recognizes that we live in the invisible present and thus cannot recognize cumulative effects.

The invisible present is our inability to stand at a given point in time and see the small, seemingly innocuous effects of our actions as they accumulate over weeks, months, and years. Obviously, we can all sense change—day becoming night, night turning into day, a hot summer changing into a cold winter, and so on. But, some people who live for a long time in one place can see longer-term events and remember the winter of the exceptionally deep snow or a summer of deadly heat.

Despite such a gift, it is a rare individual who can sense, with any degree of precision, the changes that occur over the decades of their lives. At this scale of time, we tend to think of the world as being in some sort of steady state (with the exception of technology), and we typically underestimate the degree to which change has occurred—such as global warming. We are unable to sense slow changes directly, and we are even more limited in our abilities to interpret the relationships of cause and effect in these changes. Hence, the subtle processes that act quietly and unobtrusively over decades reside cloaked in the invisible present, such as gradual declines in habitat quality.

At length, however, cumulative effects, gathering themselves below our level of conscious awareness, suddenly become visible. By then, it is too late to retract our decisions and actions even if the outcome they cause is decidedly negative with respect to our intentions. So it is that cumulative effects from our activities multiply unnoticed until something in the environment shifts dramatically enough for us to see the outcome through casual observation. That shift is defined by a threshold of tolerance in the system, beyond which the system as we knew it, suddenly, visibly, becomes something else. Within our world, this same dynamic takes place in a vast array of scales in all natural and artificial systems, from the infinitesimal to the gigantic.

At a personal level, everyone experiences cumulative effects, lag periods, and thresholds when they become ill, even if it is just a common cold. For instance, if you go to a social function, you may become infected with the cold virus, something you would not know. In fact, you would be unaware of the virus now multiplying in your body, a phenomenon that may continue unnoticed for some days (the cumulative effects within the lag period, or in parlance of disease, the *incubation period*). At length, you begin to sense something is wrong; you just do not feel "up to snuff" (the threshold); and shortly thereafter, you have the full-blown symptoms of the classic cold. In this case, the entire process encompasses a few days—from infection to expression.

A shorter-term example of cumulative effects, lag period, and threshold is the cutting down of my neighbor's dying walnut tree. Initially, a man from the tree service sawed off the small branches with intact twigs. The effect was barely discernable at first, even as they began to pile up on the ground. Each severed branch represented a cumulative effect that would have been all but unnoticeable had they not been accumulating under the tree.

After an hour or so (lag period) of removing the small limbs on one side of the tree, the cumulative effects gradually became visible as they crossed the threshold. Had the same volume of twigs been removed from throughout the tree and simultaneously gathered and removed from the ground, the cumulative effects would not have been as apparent. Nevertheless, the tree was gradually transformed into a stark skeleton of larger branches and the main trunk. Then the large branches were cut off a section at a time, with the same visual effect as when the small ones had been removed, until only the trunk remained. The piecemeal removal of the tree created a slowly changing vista of my neighbor's house, until I had an unobstructed view of it for the first time, as another stark threshold was crossed.

If we now increase the spatial magnitude that encompasses the formation of a river's delta, the timescale involved for the cumulative effects to cross the threshold of visibility may well require centuries to millennia. When a river reaches the sea, it slows and drops its load of sediment. As the amount of sediment accrues on the seabed, it diverts the river's flow, causing it to deposit additional sediment loads in other areas (cumulative effects). Thus, over many years (lag period), the accumulated sediment begins to show above the water (threshold) and increasingly affects the river's flow as it forms a classic delta. The speed with which the delta grows has numerous variables, such as the amount of precipitation within the river's drainage basin in any given year as well as the amount of its annual sediment load. Many of today's extant river deltas began developing around 8,500 years ago, as the global level of the seas stabilized following the end of the last ice age.[36] And so the process of change and novelty continues unabated in all its myriad and astounding scales.

Principle 12: All Systems Are Cyclical, But None Is a Perfect Circle

While all things in Nature are cyclical, no cycle is a perfect circle, despite such depictions in the scientific literature and textbooks. They are, instead, a coming together in time and space at a specific point, where one "end" of a cycle approximates—*but only approximates*—its "beginning" in a particular time and place. Between its beginning and its ending, a cycle can have any configuration of cosmic happenstance. Biophysical cycles can thus be likened to a coiled spring insofar as every coil approximates the curvature of its neighbor but always on a different spatial level (temporal level in Nature), thus never touching.

The size and relative flexibility of a metal spring determines how closely one coil approaches another—the small, flexible, coiled spring in a ballpoint pen juxtaposed to the large, stiff, coiled spring on the front axel of an eighteen-wheel truck. The smaller and more flexible a spring, the closer are its coils, like the cycles of annual plants in a backyard garden or a mountain meadow. Conversely, the larger and more rigid a spring, the more distant are its coils from one another, like the millennial cycles of Great Basin bristlecone pines growing on rocky slopes in the mountains of Nevada, where they are largely protected from fire, or a Norway spruce growing on a rocky promontory in the Alps of Switzerland.

Regardless of its size or flexibility, a spring's coils are forever reaching outward. With respect to Nature's biophysical cycles, they are forever moving toward the next level of novelty in the creative process and so are perpetually embracing the uncertainty of future conditions—never to repeat the exact outcome of an event as it once happened. This phenomenon occurs even in times of relative climatic stability. Be that as it may, progressive global warming will only intensify the uncertainties.

In human terms, life is composed of rhythms or routines that follow the cycles of the universe, from the minute to the infinite. We humans most commonly experience the nature of cycles in our pilgrimage through the days, months, and years of our lives wherein certain events are repetitive—day and night, the waxing and waning of the moon, the march of the seasons, and the coming and going of birthdays, all marking the circular passage we perceive as time within the curvature of space. In addition to the visible manifestation of these repetitive cycles, Nature's biophysical processes are cyclical in various scales of time and space, a phenomenon that means all relationships are simultaneously cyclical in their outworking and forever novel in their outcomes.

Some cycles revolve frequently enough to be well known in a person's lifetime, like the winter solstice. Others are completed only in the collective lifetimes of several generations, like the life cycle of a three-thousand-year-old giant sequoia in California's Sequoia National Park—hence the notion of the invisible present. Still others are so vast that their motion can only be assumed. Yet, even they are not completely aloof because we are kept in touch with them through our interrelatedness and interdependence. Regarding cycles, farmer and author Wendell Berry said, "It is only in the processes of the natural world, and in analogous and related processes of human culture, that the new may grow usefully old, and the old be made new."[37]

Principle 13: Systemic Change Is Based on Self-Organized Criticality

When dealing with scale (a small, mountain lake as opposed to the drainage basin of a large river, such as the Mississippi in the United States or the Ganges in India), scientists have traditionally analyzed large, interactive systems in the same way that they have studied small, orderly systems, mainly because their methods of study have proven so successful. The prevailing wisdom has been that the behavior of a large, complicated system could be predicted by studying its elements separately and by analyzing its microscopic mechanisms individually—the reductionist-mechanical thinking predominant in Western society that tends to view the world and all it contains through a lens of intellectual isolation. During the last few decades, however, it has become increasingly clear that many complicated systems, like forests, oceans, and even cities do not yield to such traditional analysis.

Instead, large, complicated, interactive systems seem to evolve naturally to a critical state in which even a minor event starts a chain reaction that can affect any number of elements in the system and can lead to a dramatic alteration in the system. Although such systems produce more minor events than catastrophic ones, chain reactions of all sizes are an integral part of system dynamics. According to the theory called *self-organized criticality*, the mechanism that leads to minor events (analogous to the drop of a pin) is the same mechanism that leads to major events (analogous to an earthquake).[38] Not understanding this, analysts have typically blamed some rare set of circumstances (some exception to the rule) or some powerful combination of mechanisms when catastrophe strikes.

Nevertheless, ecosystems move inevitably toward a critical state, one that alters the ecosystem in some dramatic way. This dynamic makes ecosystems dissipative structures in that energy is built up through time only to be released in a disturbance of some kind, such as a fire, flood, or landslide; in some scale, ranging from a freshet in a stream to the eruption of a volcano; after which energy begins building again toward the next release of pent-up energy somewhere in time.

Such disturbances, as ecologists think of these events, can be long term and chronic, such as large movements of soil that take place over hundreds of years (termed an *earth flow*), or acute, such as the crescendo of a volcanic eruption that sends a pyroclastic flow sweeping down its side at amazing speed. (A *pyroclastic flow* is a turbulent mixture of hot gas and fragments of rock, such as pumice, that is violently ejected from a fissure and moves with great speed down the side of a volcano. *Pyroclastic* is Greek for "fire-broken.")

Here, you might interject that neither a movement of soil nor a volcano is a living system in the classical sense. Although that is true, all disturbance regimes are part and parcel of the living systems they affect. Thus, interactive systems, from the habitat of a gnat to a tropical rainforest, perpetually organize themselves to a critical state in which a minor event can start a chain reaction that leads to a catastrophic event—as far as living things are concerned, after which the system begins organizing itself toward the next critical state. Furthermore, such systems never reach a state of equilibrium but rather evolve from one semistable state to another. This dynamic is precisely why sustainability is a moving target—not a fixed endpoint or a steady state.

Principle 14: Dynamic Disequilibrium Rules All Systems

If change is a universal constant in which nothing is static, what is a natural state? In answering this question, it becomes apparent that the *balance of Nature* in the classical sense (disturb Nature and Nature will return to its former state after the disturbance is removed) does not hold. In fact, the so-called balance of Nature is a romanticized figment of the human imagination, something we conjured to fit our snapshot image of the world in which we live. In reality, Nature exists in a continual state of ever-shifting disequilibrium wherein ecosystems are entrained in the irreversible process of change, thereby altering their composition, interactive feedback loops, and thus the use of available resources—irrespective of human influence. Perhaps the most outstanding evidence that an ecosystem is subject to constant change and disruption rather than remaining in a static balance comes from studies of naturally occurring external factors that dislocate ecosystems, and climate appears to be foremost among these factors.

After a fire, earthquake, volcanic eruption, flood, hurricane, or landslide, for example, a biological system may eventually be able to approximate what it was through resilience—the ability of the system to retain the integrity of its basic relationships. But, regardless of how closely an ecosystem might approximate its former state following a disturbance, the existence of every ecosystem is a tenuous balancing act because every system is in a continual state of reorganization that occurs over various scales of time, from the cycle of an old forest to the geological history of Zion National Park in the state of Utah.

Bear in mind, an old forest that is burned, blown over in a hurricane, or smashed in a tsunami could be replaced by another, albeit different, old forest on the same acreage. In this way, despite a repetitive disturbance regime, a forest ecosystem can remain a forest ecosystem. Thus, ancient forests around the world have been evolving from one critical state to the next, from one natural catastrophe to the next.

On the other hand, formation of the canyon in Zion National Park has a much longer history than any of the world's forest. Where today the deep canyons and massive walls of stone enthrall visitors, 245 million years ago a sea covered the area that was populated by marine fishes. Over a period of roughly 35 million years, about 1,800 feet (549 meters) of sediments were deposited on the floor of the sea, along the coastal plain, and along the inland streams.

As the climate warmed, the sea changed into a gigantic swamp. Here, 210 million years ago, crocodile-like, plant-eating dinosaurs swam in the sluggish streams whose floods carried drifted trees on their swirling waters from distant forests to form logjams. Here, also, small, fragile dinosaurs hunted along the banks of the streams. But, as the climate once again became moister during the next 40 million years, the swamp became a lake, and the sand, silt, and clayey mud of the streams and the swamp gradually hardened into rock.

The lake for a time had fish living in it, but then some of its waters became shallow and eventually disappeared. And, existing streams spread silt and sandy mud over the sediments deposited on the lake's bottom. Toward the end of this forty-million-year interval, the climate began to dry, and in a short space of time, geologically speaking, the now-intermittent streams deposited more sediments.

Then, about 170 million years ago, the ancient sea, the swamp, the lake, and the intermittent streams became buried beneath a desert of marching sand dunes. This now-hostile environment had little life associated with it, and the few hardy plants and animals that did exist often died during the great storms that blew clouds of hot, dry sand into dunes. As the dunes were built, destroyed, and built again, some of the plants and animals became entombed and are the rare fossils of today in what is now the sandstone, ranging from 1,500 to 2,000 feet (457 to 610 meters) thick. Although the source of the sand eroded away 150 million years ago, evidence indicates that it had been a region of highlands in what is today the state of Nevada.

For a brief period following the creation of the desert, floodwaters, carrying suspended sediments, buried the dunes in deposits of red mud, after which the climate returned to more desertlike conditions.

Again, the climate changed, and 145 million years ago a vast, shallow sea once more covered the area, drowning the desert. Now, the once-sterile desert, with its cap of red mud, became the floor of the sea and the home of sea lilies (crinoids) and of shellfish. But, when the warm, teeming waters once again retreated, they left behind, buried in limey silt, shells that produced the present-day fossils.

Over the millions of years, in response to changing environmental conditions, various materials were deposited in the sediments. The Zion area experienced shallow seas, coastal plains, a giant swamp, a lake, intermittent streams, and a desert filled with massive, windblown dunes of sand. While the shallow seas covered the area, mineral-laden waters slowly filtered down through the layers of sediment. Minerals like iron and calcium carbonate were deposited in the spaces between the particles of silt, sand, and mud, cementing them together, thereby turning them into stone. And, the weight of each layer caused the basin to sink and maintained its surface at an elevation near sea level. This process of deposition-sinking-deposition-sinking continued layer on layer until the accumulation of the successive sediments became 10,000 feet (3,048 meters) thick.

Geologists believe that Zion was a relatively flat basin with an elevation near sea level from 245 million years ago until the last shallow sea dried, about 10 million years ago. At that time, Zion was a featureless plain across which streams meandered lazily as they dropped their loads of sediment in sandbars and floodplains.

Then, in an area extending from Zion to the Rocky Mountains, a massive geologic event began. Forces deep within the Earth's mantle started to push upward on the surface of the Earth. The land in Zion rose from near sea level to as much as 10,000 feet (3,048 meters) above sea level.

Zion's location on the western edge of the uplift caused the streams to tumble off the Colorado plateau, flowing rapidly down a steep gradient. The Virgin River is illustrative because it drops more than 4,000 feet (1,219 meters) from the northeast corner of Zion National Park in Utah to Lake Mead in Arizona, 145 miles away; in comparison, the upper Mississippi River drops only 210 feet (64 meters) from Lake Itasca, in the state of Minnesota, to Grand Rapids, also in the state of Minnesota, also a distance of 145 miles.

And, because fast-flowing water carries more sediments and larger boulders than does slow-moving water, these swift streams in Zion began eroding down into the layers of rock, cutting deep, narrow canyons. In the ten thousand years since the uplift began, the North Fork of the Virgin River has both carved Zion Canyon and carried away a layer of rock nearly 5,000 feet (1,524 meters) thick, a layer that once lay above the highest existing rock in the park.

The uplift of the land is still occurring, so the Virgin River is still excavating. The river, with its load of sand, has been likened to an ever-moving strip of sandpaper. Its grating effect, coupled with the steepness of the Colorado Plateau, has allowed the river to cut its way through the sandstone in a short time, geologically speaking.

The cutting of Zion Canyon created a gap in the solid layer of resistant sandstone, and the walls of the canyon relaxed and expanded ever so slightly toward this opening. Because rock is generally rigid, this expansion caused cracks, known as *pressure-release joints*, to form inside the canyon's walls. These cracks run parallel to the canyon about 15 to 30 feet (4.6 to 9.0 meters) inside the walls and occur throughout the sandstone.

The grains of sand that form the sandstone itself were once driven bouncing across the desert by the wind, only to be caught within the steep face of a dune, where they became buried. Over time, the cement of lime tied grain to grain, creating the stone of sand.

That process is now reversed, and a new cycle has begun. The layer of siltstone directly beneath the sandstone is softer and more easily eroded than the sandstone. Thus, as the walls of sandstone are undermined by the erosion of this softer material, water from rain and snow seeps into the joints, where it freezes in winter, wedging the walls of the joints ever further apart.

In addition to freezing, the water, one drop of rain at a time, one melting flake of snow at a time, aided by chemical action, dissolves the cement. The structure gradually weakens, until a last grain of sand holding the undermined wall in place moves, and the massive piece of rock falls. Breaking away along the line of least resistance, it leaves the graceful sweep of a huge arch sculpted in the face of the cliff at 1,000 feet (305 meters) above the floor of the canyon. And so is revealed yet another vertical face previously hidden as a crack or pressure-release joint inside the wall. Below, the rock, shattered by the fall, gradually returns to sand and is once again blown hither and yon by the wind or carried toward the sea by the restless Virgin River.

In the end, Zion, cemented together grain by grain over millions of years, is being dissolved over millions of years one grain at a time by the persistence of water from rain and melting snow. But, while Zion undergoes its inevitable changes, it is the home for 670 species of flowering plants and ferns, 30 species of amphibians and reptiles, 125 species of resident birds, and 95 species of mammals. Nevertheless, the wolf, grizzly bear, and native bighorn sheep are gone, extirpated within the last 150 years or so by the invading European American settlers. Thus are tipped once again the scales of disequilibrium in all its dimensions.[39]

Summation

Chapter 3 examined the inviolate biophysical principles that regulate how our world functions. These principles are the active constraints that inform us about the limitations Nature places on our human ambitions—limitations we must voluntarily account for in all aspects of social-environmental planning if we are truly committed to long-term sustainability for all generations. If we continue to ignore—or blatantly refuse to heed—Nature's biophysical principles, our lack of leadership will commit all generations to pay for our lack of social-environmental consciousness. Chapter 4 is the application of these principles through the medium of design.

Notes

1. Carl Sagan. First Science.com. http://www.firstscience.com/home/poems-and-quotes/quotes/carl-sagan-quote_2284.html (accessed January 2, 2009).
2. *The Holy Bible*. Genesis 1:2. Authorized King James Version. World Bible Publishers, Iowa Falls, IA.
3. Stanford Linear Accelerator Center. The Virtual Visitor Center of Stanford University. http://www2.slac.stanford.edu/vvc/theory/quarks.html (accessed December 9, 2008).
4. Rod Nave. Department of Physics and astronomy, Georgia State University, Atlanta. Quarks. http://hyperphysics.phy-astr.gsu.edu/hbase/Particles/quark.html (accessed December 9, 2008).
5. Rod Swenson. Emergent evolution and the global attractor: The evolutionary epistemology of entropy production maximization. *Proceedings of the 33rd Annual Meeting of The International Society for the Systems Sciences*, P. Leddington (ed.)., 33(3), 46–53, 1989; Rod Swenson. Order, evolution, and natural law: Fundamental relations in complex system theory. In: *Cybernetics and Applied Systems*, C. Negoita (ed.), 125–148. Marcel Dekker Inc., New York, 1991.
6. Rod Swenson and Michael T. Turvey. Thermodynamic reasons for perception-action cycles. *Ecological Psychology*, 3 (1991):317–348.
7. Rod Swenson. Spontaneous order, autocatakinetic closure, and the development of space-time. *Annals New York Academy of Sciences*, 901 (2000):311–319.

8. Marcus Aurelius. BrainyQuote. http://www.brainyquote.com/quotes/authors/m/marcus_aurelius.html (accessed December 30, 2008).
9. G. O. Poinar, A. E. Treat, and R. V. Southeott. Mite parasitism of moths: Examples of paleosymbiosis in Dominican amber. *Experientia*, 47 (1991):210–212.
10. The general discussion of amberization is based on George O. Poinar Jr. Insects in amber. *Annual Review of Entomology*, 46 (1993):145–159; Anonymous. Scientist: Earth's oldest spider web discovered. London. *Corvallis Gazette-Times* (December 16, 2008); Enrique Peñalver, David A. Grimaldi, and Xavier Delclòs. Early cretaceous spider web with its prey. *Science*, 312 (2006):1761.
11. Chris Maser. *Earth in Our Care: Ecology, Economy, and Sustainability*. Rutgers University Press, Piscataway, NJ (2009).
12. Paul May. Chlorophyll. http://www.chm.bris.ac.uk/3motm/chlorophyll/chlorophyll_h.htm (accessed January 5, 2009).
13. Maser. *Earth in Our Care*.
14. The discussion of elk is based in part on Jack Ward Thomas, Hugh Black Jr., Richard J. Scherzinger, and Richard J. Pederson. Deer and Elk. In: *Wildlife Habitats in Managed Forests: The Blue Mountains of Oregon and Washington*. Jack Ward Thomas (technical ed.), 104–127. U.S. Department of Agriculture, Forest Service, Pacific Northwest Range and Experiment Station, Portland, OR, 1979. Agricultural Handbook No. 553.
15. Hiroki Hata and Makoto Kato. A novel obligate cultivation mutualism between damselfish and polysiphonia algae. *Biology Letters*, 2 (2006):593–596; Hiroki Hata and Makoto Kato. Monoculture and mixed-species algal farms on a coral reef are maintained through intensive and extensive management by damselfishes. *Journal of Experimental Marine Biology and Ecology*, 313 (2004):285–296; Hiroki Hata and Makoto Kato. Weeding by the herbivorous damselfish *Stegastes nigricans* in monocultural algae farms. *Marine Ecology Progress Series*, 237 (2002):227–231; Hiroki Hata and Makoto Kato. Demise of monocultural algal farms by exclusion of territorial damselfish. *Marine Ecology Progress Series*, 263 (2003):159–167.
16. Hiroki Hata, Moritaka Nishihira, and S. Kamura. Effects of habitat-conditioning by the damselfish *Stegastes nigricans* on community structure of benthic algae. *Journal of Experimental Marine Biology and Ecology*, 280 (2002):95–116; Hiroki Hata and Moritaka Nishihira. Territorial damselfish enhances multi-species co-existence of foraminifera mediated by biotic habitat structuring. *Journal of Experimental Marine Biology and Ecology*, 270 (2002):215–240.
17. David Archer. Carbon cycle: Checking the thermostat. *Nature Geoscience*, 1 (2008):289–290.
18. Richard E. Zeebe and Ken Caldeira. Close mass balance of long-term carbon fluxes from ice-core CO_2 and ocean chemistry records. *Nature Geoscience*, 1 (2008):312–315.
19. The discussion of color changes in buildings is based on Carlotta M. Grossi, Peter Brimblecombe, Rosa M. Esbert, and Francisco Javier Alonso. Color changes in architectural limestone from pollution and cleaning. *Color Research and Application*, 32 (2007):320–331; Catherine Brahic. Cleaner air to turn iconic buildings green: With atmospheric changes, limestone buildings will turn yellow, reddish-brown and green. *New Scientist*, http://www.newscientist.com/article/dn16198-cleaner-air-to-turn-iconic-buildings-green.html (accessed December 8, 2008).

20. C. I. Davidson, W. Tang, S. Finger, V. Etyemezian, M. F. Striegel, and S. I. Sherwood. Soiling patterns on a tall limestone building: Changes over sixty years. *Environmental Science and Technology*, 34 (2000):560–565.
21. Cliff I. Davidson. Air pollution in Pittsburgh: A historical perspective. *Journal of the Air Pollution Control Association*, 29 (1979):1035–1041.
22. V. Etyemezian, C. I. Davidson, M. Zufall, W. Dai, S Finger, and M. Striegel. Impingement of rain drops on a tall building. *Atmospheric Environment*, 34 (2000):2399–2412; Vicken Etymezian, Cliff I. Davidson, Susan Finger, and others. Vertical gradients of pollutant concentrations and deposition fluxes at the cathedral of learning. *Journal of the American Institute for Conservation*, 37 (1998):187–210.
23. Carlotta M. Grossi, Peter Brimblecombe, Rosa M. Esbert, and Francisco Javier Alonso. Color changes in architectural limestone from pollution and cleaning. *Color Research and Application*, 32 (2007):320–331; Alessandra Bonazza, Peter Brimblecombe, Carlota M. Grossi, and Cristina Sabbioni. Carbon in black crusts from the Tower of London. *Environmental Science and Technology*, 41 (2007):4199–4204.
24. Grossi et al. Color changes.
25. Grossi et al. Color changes; Bonazza et al. Carbon in black crusts.
26. Bill Zlatos. Foes, lack of funds may scrub cathedral of learning cleaning. *Pittsburgh Tribune-Review* (July 6, 2003).
27. Sid Perkins. As ozone hole heals, Antarctic could heat up. *Science News* (July 5, 2008):10; S.-W. Son, L. M. Polvani, D. W. Waugh, and others. The impact of stratospheric ozone recovery on the Southern Hemisphere westerly jet. *Science*, 320 (2008):1486–1489; J. Perlwitz, S. Pawson, R. L. Fogt, and others. Impact of stratospheric ozone hole recovery on Antarctic climate. *Geophysical Research Letters*, 35 (2008):L08714, doi:10.1029/2008GL033317 (accessed December 17, 2008).
28. Lee Dye. Did climate change kill the Roman Empire? http://abcnews.go.com/Technology/JustOneThing/story?id=6428550&page=1 (accessed December 10, 2008).
29. Sid Perkins. Buried treasures. *Science News*, 169 (2006):266–268; Martin B. Short, James C. Baygents, and Raymond E. Goldstein. Stalactite growth as a free-boundary problem. *Physics of Fluids*, 17 (2005):083101, 12 pages. http://www.math.ucla.edu/~mbshort/papers/stalactite2.pdf (accessed December 17, 2008); M. B. Short, J. C. Baygents, J. W. Beck, and others. Stalactite growth as a free-boundary problem: A geometric law and its platonic ideal. *Physical Review Letters*, 94 (2005):018510, 4 pages. http://www.math.ucla.edu/~mbshort/papers/stalactite1.pdf (accessed December 17, 2008).
30. D. Meko, C. A. Woodhouse, C. A. Baisan, and others. Medieval drought in the upper Colorado River Basin. *Geophysical Research Letters*, 34 (2007): L10705, doi:10.1029/2007GL029988. http://www.agu.org/pubs/crossref/2007.../2007GL029988.shtml (accessed December 17, 2008).
31. Ian J. Orland, Miryam Bar-Matthews, Noriko T. Kita, and others. Climate deterioration in the eastern Mediterranean as revealed by ion microprobe analysis of a speleothem that grew from 2.2 to 0.9 Ka in Soreq Cave, Israel. *Quaternary Research*, 71 (2009):27–35; A. Kaufman, G. J. Wasserburg, D. Porcelli, and others. U-Th isotope systematics from the Soreq Cave, Israel and climatic correlations. *Earth and Planetary Science Letters*, 156 (1998):141–155; Avner Ayalon,

Miryam Bar-Matthews, and Eytan Sass. Rainfall-recharge relationships within a karstic terrain in the eastern Mediterranean semi-arid region, Israel: $\delta\,^{18}O$ and δD Characteristics. *Journal of Hydrology*, 207 (1998):18–31.
32. Pingzhong Zhang, Hai Cheng, R. Lawrence Edwards, and others. A test of climate, sun, and culture relationships from an 1810-year Chinese cave record. *Science*, 322 (2008): 940–942; Kallie Szczepanski. When the Rains Stop, the Emperors Fall. http://asianhistory.about.com/od/asianenvironmentalhistory/a/ChinaMonsoon.htm (accessed January 5, 2009); Ker Than. Chinese kingdoms rose, fell with monsoons? *National Geographic News*, http://news.nationalgeographic.com/news/2008/11/081106-monsoons-china.html (accessed January 10, 2009); Yongjin Wang, Hai Cheng, R. Lawrence Edwards, and others. Millennial- and orbital-scale changes in the East Asian monsoon over the past 224,000 years. *Nature*, 451 (2008):1090–1093.
33. Charles Darwin. *On the Origin of Species*. Modern Library, a Division of Random House Publishers, New York (1998).
34. Maser. *Earth in Our Care*.
35. Francis Bacon. http://Science.prodos.org (accessed January 2, 2009).
36. Sid Perkins. O river deltas, where art thou? *Science News*, 172 (2007):118; Pippa L. Whitehouse, Mark B. Allen, and Glenn A. Milne. Glacial isostatic adjustment as a control on coastal processes: An example from the Siberian Arctic. *Geology*, 35 (2007):747–750.
37. Wendell Berry. The road and the wheel. *Earth Ethics*, 1 (1990):8–9.
38. Per Bak and Kan Chen. Self-organizing criticality. *Scientific American* (January 1991):46–53.
39. The foregoing discussion of Zion National Park is based on A. J. Eardley and James W. Schaack. *Zion: The Story Behind the Scenery*. KC Publications Inc., Las Vegas, NV (1989).

4

Basic Components of Design

Introduction

Design, in the context of this book, is the organization of elements in the sense of composition. *Composition,* in turn, is the combination of parts or elements to form a systemic whole. And, a *systemic whole* of any kind has five basic components: (1) composition, (2) structure, (3) function, (4) systems-altering disturbances, and (5) constraints with varying levels of negotiability. In this book, I contrast a forest with a city because every city is composed of design elements taken from Nature (largely from forests) and is best comprehended in that way if we are to successfully adapt to the vagaries of global climate change.

Composition

Everyforest

In a forest, composition consists of the number and kinds of plants that grow in a particular area and the length of time they live—coniferous trees, deciduous trees, shrubs, grasses, forbs, mosses, and lichens, as well as fungi that, strictly speaking, are not plants. The longevity of a particular kind of plant is critical because that is the length of time the plant affects the site whereon it grows.

Further, trees may grow relatively close together in dense stands or widely spaced in mountain meadows. They may grow large along streams and stunted on rocky ridge tops or in poor soil. Tall trees dominate smaller trees and thus garner most of the sun's light, while the smaller ones, shrubs of various kinds, and numerous varieties of herbaceous (nonwoody) plants growing relatively close to the ground have to tolerate various degrees of shade. In this way, the collective composition of the various species creates the composition of a forest.

Everycity

In addition to a city's vegetation and its extant natural features, such as a river or rocky hill, it is the number and kinds of buildings that act in concert to create its particular compositional diversity—homes, apartments, gas stations, stores, churches, taverns, hospitals, mortuary, fire station, police station, and so on. These can be situated in areas of mixed use, like many European cities, or separated as housing developments, strip malls, industrial parks, and so on.

Moreover, the different kinds and longevity of buildings affect the areas in which they are situated. A newly constructed housing development will affect the area it occupies much differently from the way a hundred-year-old church affects the site on which it rests.

Structure

Everyforest

In a forest "system," structure is an outcome of the composition of plants that grow in an area because each plant and each kind of plant grows differently—some are tall, thin, delicately branched, evergreen, and have soft wood; others are short, stout, deciduous, robustly branched, and have hard wood. As well, the area may have a midstory of shrubs and a ground cover of mosses, forbs, grasses, or a combination of these nonwood plants. The cumulative effect of how they grow creates the vegetative structure of the area.

Everycity

In addition to a city's vegetation and extant natural features, its buildings create structural diversity because some are short, small, and broad, whereas others are tall and slender. Some are made primarily of wood, others of steel and concrete, and still others of brick or stucco. A number of buildings may be rounded, others square, and a few with a variety of angular shapes. Some are sheer sided, some have ledges and overhangs, whereas others are layered—much like geological formations found in various landscapes, including cliff faces in forests. In addition, some buildings have attics with openings that allow access to and from the outside, much like large cavities in snags or caves.[1] There are buildings with shake roofs, slate roofs, tile roofs, and composition roofs. Some roofs are pitched at different angles; others are flat.

As the composition of plants affects the kind of forest that occupies a given area, so the composition of buildings acts in accordance to affect a given city. To illustrate, Washington, D.C., is replete with the stately offices of government (local, state, and federal). Washington was consciously planned to represent moral responsibility, whereas Las Vegas, Nevada, with its plethora of gaudy casinos and other centers of entertainment, grew to be the epitome of distraction and escape. In contrast, Toulouse, France, is an architectural archive of history with its narrow streets from the days of horse and buggy and interesting old buildings. Regardless of its cultural legacy, structural components coalesce to create its functional dynamic as a system.

Function

Everyforest

The combined features of composition and structure allow certain functions to take place within a given area of the forest:

- A tree with rough, craggy bark offers numerous crevices in which small bats sleep during the day, whereas the stiff, well-clothed branches make excellent sites for a variety of birds to nest.
- A tree with a spirelike shape of stiff, rather short branches, while not particularly conducive for nesting by a wide variety of forest birds, readily withstands wind and easily sheds the snows of winter.
- A tree with rather smooth, finely textured bark and lacy branches holds little value as wildlife habitat, but if it has a relatively short life span, it creates a fairly steady supply of large snags (standing, dead trees) for cavity-nesting birds, such as woodpeckers, and cavity-nesting mammals, such as the Australian common brushtail possum. Snags with loose, but attached, slabs of bark offer roosting sites for various bats.
- Some cedar trees have much-enlarged stems at ground level; these frequently become hollow in old trees due to rot. These hollows are ideal winter dens for hibernating black bears, especially for females because they gave birth to their cubs during hibernation. In addition, large cedars often become hollow snags when they die because of a long infestation of heart rot. These hollow snags are critical nesting habitat for swifts, small birds that fly into the hollow snag from the top and fasten their nest to the inside wall of the dead tree.
- Spruce trees often have relatively droopy branches well enough covered in stiff needles that some birds can use them for nesting.

Everycity

In addition to a city's vegetation and extant natural features, its buildings, bridges, and other structures create a particular functional diversity. Between a city's vegetation and its human-made structures, virtually every function of habitat found in a forest can be correlated to a similar function within a city.

- As a tree's rough, craggy bark offers crevices in which small bats can sleep during the day, bats use the crevices in shake roofs and behind window shutters. Other bats use the crevices in tile roofs and the undersides of concrete bridges.
- Buildings with openings to the interior of outer walls or other areas often emulate tree cavities for such cavity-nesting birds as starlings.
- In some cities, ledges near the top of tall buildings are used as nesting platforms by falcons.
- Chimneys serve in a fashion similar to large, hollow snags for nesting swifts—an important attribute because the requisite snags are fast disappearing from most forests.
- The undersides of bridges emulate the overhangs and ledges of cliff faces that accommodate part of the habitat requirements of domestic pigeons. In some cases, bridges also contain crevices and cavelike structures that bats use for roosting and rearing their young, as well as surfaces that cliff swallows use for attaching their mud nests.[2]
- Houses with attics that have access to the outside sometimes act as hollow trees in which bats form nursery colonies. In other instances, house mice, roof rats, and squirrels make their abodes in available attics.
- Bats roost under the loose siding of homes, much as they do under the loose bark on snags in a forest.

In many areas, forests that once bordered rural communities have been converted to plantation-like tree farms. By way of example, while in western Germany in 1985, I observed that plantation forests had been so simplified through human use over the centuries that a vastly richer diversity of wildlife was found within a city's limits than in the surrounding countryside. This seemed especially true for birds that used urban habitats that emulated those once found in the forests of old.

Forest-City Interface

Where along the continuum of naturalness a particular forest is situated depends on the ecological integrity of its biological composition, structure, and function just as the viability of a city's infrastructure depends on the integrity of its composition, structure, and function. The problem with the myriad relationships among composition, structure, and function is that people perceive objects in terms of their obvious structures or functions but seldom understand the role that composition plays in either.

To illustrate, a timber company clear-cuts a forest and thereby changes the plant community's composition from predominantly trees to all herbaceous plants. This shift in composition alters how the community functions because a dead stalk of grass is clearly not substitutable for a large snag when it comes to available habitat for a woodpecker. The same is true for a city in which an old, mixed-use area is razed and converted into an economically designed shopping mall, a move that excludes a prospective homeowner from finding a permanent abode in this same space.

Composition, if you remember, is the combining of a variety of parts or elements to form a whole. In the case of a forest, composition refers to the kinds of plants that compose its basic living parts. Structure is the configuration of those parts or elements, be it simple or complex. Structure can be thought of as the organization or arrangement of the plants that form the living foundation of the forest. Function, on the other hand, is what a particular structure either can do or allows to be done. So, the composition of a forest creates its structure, and the structure determines how the forest functions. Conversely, how a forest functions dictates its necessary structure, and the structure in turn dictates its necessary composition.

The same scenario is true for a city, wherein composition refers to the type and arrangement of its buildings and other human-created structures, such as bridges, streets, sewers, and so on. The *overall* structure is the configuration of those parts or elements, be it simple or complex. Structure, therefore, can be thought of as the organization or arrangement of the buildings, bridges, and other architectural elements that form the livable foundation of the city. Function, on the other hand, is what a particular structure either can do or allows to be done within the city. So, the composition of a city creates its overall structure, and the structure determines how the city functions. Conversely, how the city functions dictates its necessary structure, and that in turn dictates the necessary composition, a dynamic that precludes the functions of Washington, D.C.; Las Vegas, Nevada; and Toulouse, France, from being interchangeable as the cities are now constituted.

In Nature, on the other hand, how smoothly a given ecosystem, such as a forest, functions over time depends in large measure on the kind and frequency of disturbance events the system is subjected to; the same is true for a city.

System-Altering Disturbances

Nature's disturbances (fires, floods, windstorms, and so on) control how ecosystem processes function, such as forest succession, and thereby alter habitat. True, we humans can tinker with Nature's disturbance regimes, such as the suppression of fire, but in the end our tinkering catches up with us, and we pay the price.

In a city, the alteration of a long-standing zoning ordinance, like fire suppression in a forest, is tantamount to tinkering with one of Nature's disturbance regimes since a zoning ordinance is intended to control how the process of land use planning takes place. Each changes circumstances and thereby alters the potential outcome, and each has a degree of irreversibility attached to it.

At this juncture, it must be clearly understood that composition is the determinant of the overall structure and function because composition is the *cause* of the structure and function rather than its effect. It needs to be understood also that a major disturbance to a system has the effect of altering composition—hence structure and function. If, for instance, a pine forest were maintained as an open forest by repeated fires, the exclusion of fire would allow a shade-tolerant tree to take over and alter the forest. In a city, where racism, such as that perpetrated by the Ku Klux Klan (KKK), neo-Nazis, or religious extremists of any kind, maintains an all-white neighborhood, the elimination of the KKK, neo-Nazis, or religious extremists would allow people of other races and persuasions to move into the neighborhood and diversify its population. Altering the population (composition) of that neighborhood will alter the way it functions.

That said, there is a caveat with respect to disturbances. Namely, Nature is impartial when it comes to the effects of a disturbance, such as a massive lightning strike on a tree and the ensuing fire, but we humans are definitely partial and so prejudiced against major disturbances. Take, for instance, the September 11, 2001, attacks on the Twin Towers as a city comparison of the aforementioned, massive lightning strike.

In the case of the lightning strike, there are obvious changes in the forest. In addition, the lives of hundreds of thousands of plants and animals are lost—something most people are not even conscious of. But when human life is at risk and lives are lost in a terrorist attack, social upheaval and retaliation follow.

The difference between plant and animal lives lost in a catastrophic event and the lives of people lost in a catastrophic event is that people are seen as concrete, knowable entities because they are like us—and we know them as individuals. But, to most people, the plants and animals in a forest are unknown abstractions. Beyond that, forests are seen by many people simply as commodities that have value only when converted into a product, such as lumber. With this view, the major concern with a forest fire is to "salvage" the timber while its commercial value can still be realized.

There are, besides the obviously cataclysmic disturbances, minute organisms (such as viruses) that can attack a forest or people unseen but with equally devastating effects. Whereas foreign terrorists slip through security and attack cities, foreign insects slip through "security" and attack forests. Although there are, as you might guess, numerous examples on which I could draw, these suffice to make the point that many such commonalties exist.

Beyond tinkering with a disturbance regime, we can change the trajectory of an ecosystem, such as a forest, by altering the kinds and arrangement of plants within it through "management" practices because that composition is malleable to human desire and, being malleable, is negotiable within the context of cause and effect. Clearly, then, the same is true for a city. If the "downtown" area of a city is revitalized and grossly altered in the process, such as by the exclusion of automobiles, it will function in a manner that is entirely different from its previous incarnation.

By changing the composition to meet a particular desire, we simultaneously change both the overall structure and the function. Once the composition is ensconced, structure and function are set on a predetermined trajectory—unless, of course, the composition is drastically altered, at which time both the structure and function are altered accordingly.

If we change the plant composition of a forest or the composition of buildings in a city, we change the structure of the system, hence the function, and that affects the animals and people. Under Nature's scenario, the animals are ultimately constrained by the composition because, once the composition is in place, the structure and its attendant functions operate as a unit in terms of the habitat opportunities that either fulfill the requirements for the various species of animals or they do not. The same is true for people in a city. Convert a slum to a middle-class housing development, and the previous tenants will be excluded in favor of people from a higher socioeconomic class, thereby altering how the overall system functions.

But then, people and Nature are continually manipulating the composition of forests and cities, thereby changing the composition of both the animal and human communities dependent on the structure and function of the resultant habitats. These manipulations in turn determine what kinds of animals and which people and how many can live where, as well as what uses humans can make of the system. Ergo, in order to maintain or restore the biological health of a forest or the socioeconomic health of a city so it can provide the things we valued it for in the first place, it is necessary to figure out how such a forest or city functions. With this knowledge in hand, one has to work backward through the required physical structure to the necessary composition to achieve a desired outcome because the latter is based on the negotiability of constraints.

The Negotiability of Constraints

To achieve a desired outcome, one is obliged to have a vision of what is wanted.[3] A vision of some future, desired condition, by its very nature, elicits constraints that must be met if terms of the vision are to be fulfilled. A constraint, in this sense, connotes something that restricts, limits, or regulates personal, human behavior.

A vision does not in and of itself create constraints where none existed. It cannot because everything in the world is already constrained by its relationship to everything else, a circumstance that means nothing is ever entirely free. The people who create the vision determine the behavioral constraints necessary to achieve a collectively chosen outcome. Moreover, a shared community vision determines the degree to which a particular behavioral constraint is negotiable.

What does *negotiable* mean here? It means to bargain for a different outcome, to cut the best possible deal. For instance, most changes in climate are determined by Nature and are nonnegotiable. Can we negotiate with Nature to give us sunnier, drier winters without flooding when we deem the winters too dark and wet? Can we negotiate for more rain during a summer we deem too dry? Well, we can try, but it would be to no avail. While somewhat flexible at times, Nature does not negotiate. Therefore, the conditions Nature hands us are, in the final analysis, *nonnegotiable*. We cannot cut a "better" deal, one more to our liking.

The challenge we humans face is to learn what is negotiable and what is not. Beyond this, necessity dictates that we learn to accept with grace what is not negotiable and learn to account for and accept responsibility for the price of what is, because negotiability is not free. When we negotiate, we trade one set of behavioral freedoms for another in that we impose a particular constraint on ourselves through a vision in order to alleviate or free up some other potential constraint in the future—the desired outcome of our vision.

Everyforest and Everycity function fully within the limits (*constraints*) imposed on them by Nature or humans. It is thus the type, scale, and duration of the alterations to the system with which we need to be concerned when contemplating the negotiability of constraints.

To this end, we must both recognize and accept that Nature's biophysical laws, such as the second law of thermodynamics, are nonnegotiable. Just as the rising of the sun and the phases of the moon are nonnegotiable, so we cannot counter the laws of gravity by sitting under an apple tree and make its falling fruit reverse course, levitate through the air, and refasten itself to its twig. The flip side of the coin is that society's laws are negotiable because people made them and so can change them. The outcome of our decisions and subsequent actions are constrained, nonetheless, by Nature's biophysical principles, which in turn create the social-environmental consequences of our anthropogenic design.

Because we create our social laws, a legislative body can negotiate the speed with which people are legally allowed to drive on a city's streets, whether patrons can smoke in taverns, what kinds of uniforms are to be worn in the military and when. With respect to an urban-forest interface, the placement of roads, city limits, and streetlights, whether to allow hunting, and so on are all negotiable constraints. But, if a city's growth encroaches too much on the habitat of coyotes or mountain lions, their lives within the city limits are not negotiable, which means controlling them is deemed necessary.

Although these depictions tend to be either/or propositions, there are degrees of negotiability, such as the amount of light pollution in a housing development along the outskirts of a moderate-size town. I say "negotiability" because (while the brilliance of the stars' light in the night sky is nonnegotiable) the number, kind, and placement of the street lights and the number of people who have and turn on outdoor lighting in a housing development at night is negotiable. The outcome of that negotiation determines how clearly and brightly stars light the night sky from our human perspective on the ground.

Here is a critical observation. We introduce thoughts, practices, substances, and technologies into the environment, and we usually think of those introductions in terms of development. Until they have been introduced, they remain negotiable, much like the Arab proverb about the spoken word: Until a word is spoken, you own it. Once spoken, it owns you.

This proverb is but saying that until we introduce something into the environment, we are in charge, and its introduction is negotiable. But once introduced, it is effectively out of our control, and its effects—good or bad—are forever nonnegotiable.[4] Whatever we introduce into the environment in the name of development will consequently determine how the environment will respond to our presence and to our cultural necessities. For this reason, it is in our social "best interest" to pay close attention to what we introduce in the environment and, once introduced, what it is likely to affect over time. After all, values—often competing values to which we give no conscious thought—shape the contours of our lives; they also raise the questions of how we go about calculating the risk of something that has never before happened and, by our reckoning, is unlikely to occur in the future.

As society develops new technology, draws on the resources of the Earth, and generates unprecedented quantities of unintended industrial products, such as toxic wastes, the question about introductions is being asked more and more frequently because we must understand, as best we can, what the effects of our activities will be on the sustainability of our environment as a functional system. Some people would throw their hands up and say such predictions are impossible, but that is not entirely true. Over the past few decades, an entire discipline, known as *risk assessment* or *risk analysis*, has been formulated around the proposition of "what if."

Assessment of risk focuses on three issues, according to retired physicist and engineer B. John Garrick: "What can go wrong? How likely is it? And what are the consequences?" The answers are given as a probability, the language of uncertainties; learning how to quantify the uncertainties is a critical part of assessing risk.

Many people do not find the probabilities reassuring, however, despite how low they may be. To many of us, no matter how low the probabilities of adverse possibilities may *appear* to be, we know how ignorant we are collectively when it comes to the synergistic effects of interdependent living systems. These doubts are not based on fear but rather on concern for the too-often-expressed sense of infallibility of our human knowledge. Every decision has a risk attached to it, and the more we can understand the risks of our proposed actions, the better off we will be.

Making people more comfortable with the probabilities, said Garrick, is a matter of changing the terms of the debate: "During my 40-plus years in the risk [assessment] business, the questions that have come to annoy me most are 'How safe is safe?' or 'How much risk is acceptable?' These are illogical questions. The only answer that makes sense is: 'It depends on the alternatives available and on the benefits to be gained by making a certain decision.'"

Yet, there is logic to these questions. They point out that the people who ask them are frightened and do not know how to frame the questions in a way that addresses their fears. In addition, they are questions of value, including things likely to be lost, such as some long-cherished, often intangible components of one's lifestyle. Such questions are far more complicated to deal with than the simple, traditional, linear questions asked and favored by scientists and engineers, who can—*at best*—measure only tangible effects.

Garrick went on to say that the best possible assessment of potential risk requires participation—negotiation—by the public, either directly or through elected representatives. "Governments and the private sector," admonished Garrick, "need to develop mechanisms to ensure this input [by the public] without letting the process get bogged down by a few people whose entire agendas may never be expressed and whose actions lead to gross mismanagement of society's resources. Those who spread false information [on all sides] need to be held accountable—especially since the consequences of their actions can cost billions of dollars." And, these falsehoods say nothing about severely altered ecosystems or the loss of the services they provide, a loss that adds to the misery of countless people—present and future. Accordingly, the question is: Who bogs down the process and why?

"If decisions involving risk are not approached rationally, they will be made on political and emotional bases, which usually is not optimal for society," counseled Garrick.[5] But then, all truly *rational* decisions involve—but are not ruled by—emotions; after all, they are the foundation of our values as human beings. Politics, on the other hand, is often a case of hidden agendas—also based on emotions—that determine who wins and who loses. Assessing risk in a formal manner, however, does provide us with a way to better understand the possible consequences of the choices we face.

Although assessing risk *requires* systems thinking, it is too often undertaken in a linear, symptomatic manner.

To illustrate, I was once retained to act as an "expert witness" for a group of people who were protesting the relentless political pressure exerted by officials from an absentee-owned energy company who were determined to situate an electricity-generating plant in the midst of a neighborhood. The company intended to pollute both the air and a local creek (Laughing Creek), which empties into a river, which empties into a lake (the municipal water supply for a large city), which ultimately empties into the Gulf of Mexico. In this case, the officials of and lawyers for the absentee-owned energy company acted strictly as *product-oriented thinkers* with a single objective—get the energy plant built, despite the desires of local residents and regardless of any and all social-environmental consequences. I, on the other hand, was engaged to act as a *systems thinker* on behalf of the citizens (present *and* future) who would be affected in unseen and progressively negative ways by the plant's operation.

As an expert witness in years past, I had learned that the most powerful defense against individuals who would foreclose options for all generations is a series of relevant questions to which they *must* respond. I say this because the one asking the questions is in charge, and the ones having to respond to the questions are the ones on the defensive. Therefore, the questions (both scientific and social) that I posed before the court are the kind I think need to be asked in every assessment of risk.

Scientific Questions

A battery of scientific questions needs to be asked if the potential biophysical effects of a given action are to be ascertained. The purpose of the questions is to assess both the possibility *and* the probability of whether certain negative environmental consequences will result from a proposed economically motivated action. Such questions are necessary because an apparently good, short-term, economic proposition may prove to harbor potential long-term detrimental environmental effects that some future generation will have to live with, pay for, and recover from. By this, I mean a good, short-term, economic decision may in fact turn out to be a bad long-term ecological decision and so a bad, long-term economic decision. And, there inevitably are delayed social-environmental costs attached to every short-term economic decision that turns sour—costs the decision makers irrevocably commit those of the future to pay.[6]

To depict the kinds of questions that need to be asked in any assessment of social-environmental risk, I list those I posed before the court with respect to the energy company's proposed facility on Laughing Creek. The following questions are important because a stream is a sinuous continuum of habitats that transports water and *all* its ingredients downstream while neither recognizing nor respecting human boundaries—as every major oil spill has abundantly demonstrated:

1. How can the "affected area" of Laughing Creek be limited to one mile downstream from the energy company's facility when that mile is an integral part of a continuous, interactive ecosystem that will, through cumulative impacts, affect all aquatic and terrestrial life that lives within it, drinks its water, uses its vegetation as food—especially during periods of drought?
2. How will a daily discharge of 290,000 gallons of 95° Fahrenheit water from the energy company's facility affect the physical configuration and stability of the Laughing Creek ecosystem as it has evolved to cope with periods of high water and low water? Has this been researched? If not, why not? If so, what are the results?
3. Will the channel become destabilized by a daily discharge of 290,000 gallons of 95° Fahrenheit water? Will the critical instream habitats formed by dead wood and sandbars be swept away? Will the stabilizing vegetation of the banks be able to retain its life and grip on the soil with so much hot water? Has this been researched? If not, why not? If so, what are the results?
4. How will altering the flow affect the aquatic life, which has evolved to cope with periods of high water and low water? Has this been researched? If not, why not? If so, what are the results?
5. Will the discharge of 290,000 gallons of 95° Fahrenheit water be in a steady stream or in pulses that increase during hours of peak electrical generation and decrease during slack times? If the discharge of water from the facility is in pulses, how will that affect the ecology of Laughing Creek—as outlined in questions 2 and 3? Has this been researched? If not, why not? If so, what are the results?
6. How will a daily discharge of 290,000 gallons of 95° Fahrenheit water affect the microplants and animals that form the basis of the food chain that feeds the aquatic invertebrates that feed the fish and frogs, which in turn feed the raccoons, herons, eagles, and so on—especially during hot weather or drought when the dissolved oxygen in the water is already low and will be further decreased by the addition of so much hot water? Under such conditions, will there be enough dissolved oxygen for the survival of the indigenous fish? Has this been researched? If not, why not? If so, what are the results?
7. Assuming that fish and amphibians can live in such hot water, an improbability, how will a daily discharge of 290,000 gallons of 95° Fahrenheit water affect the survival of their eggs? Has this been researched and addressed? If not, why not? If so, what are the results?

8. What will be the cumulative effects of the chemicals dumped into the Laughing Creek ecosystem from a continual daily discharge of 290,000 gallons of polluted water—especially during drought when everything in the water of Laughing Creek concentrates into a small per-unit area, and wildlife come to drink the water? Has this been researched? If not, why not? If so, what are the results?
9. Where in a drought-stricken Laughing Creek ecosystem will the chemical pollutants concentrate—in the clay of the bottom and the banks? In the vegetation that uses the water for survival? In the aquatic and terrestrial animals, including livestock, that use the vegetation or water for survival? Has this been researched? If not, why not? If so, what are the results?
10. Water is one of the vital ingredients of life and draws animals, such as deer, livestock, birds, and bats, as well as many other species, from long distances—often more than a mile's radius, especially during drought—because there is no substitute for water. This necessity for water can greatly overtax a stream system as small as Laughing Creek during times of drought, virtually ensuring that all the water available to the plants and animals would be hot and polluted. How would this circumstance affect the Laughing Creek ecosystem and the livestock and wildlife (including amphibians, reptiles, and fish) that depend on it for fresh, healthy water to sustain life? Has this been researched? If not, why not? If so, what are the results?
11. Would any of the discharged chemical compounds from the energy company's facility cause diarrhea in deer, cattle, and other mammals if they were to drink the discharged water in Laughing Creek? If so, could the loss of bodily fluids cause such dehydration in already-thirsty animals to the point of killing them—especially during a drought, when all they would have available to stave off dehydration would be the hot, polluted water of Laughing Creek? Has this been researched? If not, why not? If so, what are the results?
12. What is the biophysical fate of the various chemical compounds discharged from the facility once they enter the aquatic ecosystem of Laughing Creek?
    a. How toxic to the ecosystem are the chemicals? Has this been researched? If not, why not? If so, what are the results?
    b. Is arsenic, one of the ingredients in most control agents, such as insecticides and rodenticides, present? Is arsenic cumulative in animals' bodies? If so, what form is it? If so, how does arsenic move upward in the ecosystem through the food chain? If so, how does it affect the food chain? Has this been researched? If not, why not? If so, what are the results?

c. How biodegradable, in fact, are the chemicals in the discharge from the facility? Has this been researched? If not, why not? If so, what are the results?

d. Have the "active" ingredients of the chemical compounds discharged from the facility been tested for their toxicity to the Laughing Creek ecosystem and its food chain? If not, why not? If so, what are the results?

e. What recombinations can and might the active ingredients make with the chemical compounds already in the Laughing Creek ecosystem? Could they become more toxic than the chemical compounds discharged in the effluent? Has this possibility been tested? If not, why not? If so, what are the results?

f. Have "inert" ingredients in the chemical compounds in the discharge from the facility been tested for their toxicity to the Laughing Creek ecosystem and its food chain? If not, why not? If so, what are the results?

g. What recombinations can and might the inert ingredients make with chemical compounds already in the Laughing Creek ecosystem? Could recombinations become more toxic than the chemical compounds discharged in the effluent? Has this possibility been tested? If not, why not? If so, what are the results?

h. How biodegradable, in fact, are the recombinations? Has this been researched? If not, why not? If so, what are the results?

i. Where in the ecosystem do the discharged chemicals from the facility accumulate—especially during a drought? Has this been researched? If not, why not? If so, what are the results?

j. What are the synergistic, biophysical effects (positive *and* negative) of the concentration of the chemicals? Has this been researched? If not, why not? If so, what are the results?

13. During floods, how far from Laughing Creek does the water go? Does it collect in low areas? How long does it stand? Do the plants in these flooded areas take up more chemical pollutants than they would otherwise? Has this been researched? If not, why not? If so, what are the results?

14. Assuming the plants of flooded areas absorb greater amounts of chemical pollutants from the discharged wastewater, how does the consumption of the contaminated vegetation affect livestock and wildlife? Has this been researched? If not, why not? If so, what are the results?

15. How far will the discharged chemical compounds be transported downstream from the facility through the Laughing Creek ecosystem? At what distance, in miles, will they cease to have a negative

effect? Has this been researched? If not, why not? If so, what are the results?

16. At their farthest detectable point:
    a. What other chemical compounds will those discharged from the energy company's facility recombine with on their journey downstream from the point of discharge—such as those released by communities along the creek? Has this been researched? If not, why not? If so, what are the results?
    b. How toxic will the recombinations be? Has this been researched? If not, why not? If so, what are the results?
    c. How will they affect the microplants and animals that form the basis of the food chain that feeds the aquatic invertebrates that feed the fish and frogs, which in turn feed the raccoons, herons, eagles, and so on—especially during drought when the water is already low and will concentrate wildlife and all pollutants into a small unit of area? Has this been researched? If not, why not? If so, what are the results?

Social Questions

The social questions people most often ask are based on legitimate human concerns and sometimes outright fears of the unknown. In my experience, people with vested economic interests often give glib, unsubstantiated assurances that all is well, and that any and all concerns are obviously emotional and thus clearly unfounded. In fact, it is surprisingly common for people who stand to gain economically from a given project to treat those citizens who voice their doubts about the social-environmental safety of the project as absurdly emotional and overreactive.

Such was the behavior of the energy company's officials and lawyers when the people who lived along Laughing Creek wanted to know the following: If the polluted water is taken up by the plants my animals eat, will the health of my livestock—and so *my* health and that of *my children*—be affected by the pollutants the company wanted to dump into Laughing Creek?

The behavior of the company's officials and lawyers was the same when the people in the city asked the following: Since Laughing Creek drains into the lake from which we get our drinking water, will the pollutants the company wants to dump into Laughing Creek affect the quality, and hence the safety, of the drinking water?

Both questions are, I think, reasonable and need answering, despite the fact that they may be asked in an emotional way based on fear. But then, human values are based on fulfilling our perceived emotional desires, which are often framed as *needs*. Such needs include the drive to complete an economic endeavor that will garner money for someone when that person's perceived emotional requirement is based on the sense of security brought to fruition by having more money.

In the last case, however, the person—who seeks economic gain despite probable, negative, social-environment outcomes—is required to be accountable for answering relevant questions concerning the potential consequences of their actions. Because such accountability is to be taken seriously, I posed the following questions before the court:

1. With long-distance transport of air pollutants and their ability to alter habitats, such as that of the endangered red cockaded woodpecker, the energy company is also altering the long-term habitat for people. Does the energy company have the people's permission—adults *and* children—to add pollution to the air, thereby altering the health of the plant communities that constitute the quality of the habitat in which wildlife and people live, thus adding to the potential irreversibility of the negative, cumulative effects that present *and* future generations must endure? Is not a good quality of habitat in which to live the inherent birthright of every human being?
2. Does the energy company have the people's permission—adults *and* children—to add pollution to the air everyone breathes, thereby irreversibly adding to the very long-term negative cumulative effects with which all plants and animals, including humans, are destined to live? After all, clean air is a global commons and therefore everyone's birthright. The commons is that part of the world and univese that is every person's "birthright." There are two kinds of commons. Some are gifts of Nature, such as clean air, pure water, fertile soil, a rainbow, northern lights, a beautiful sunset, or a tree growing in the middle of a village; others are the collective product of human creativity, such as the town well from which everyone draws water or a museum of fine art.
3. Does the energy company have the people's permission—adults *and* children—to add pollution to the air that can exacerbate global warming and alter the local pattern of precipitation, thus adding to the negative, cumulative effects with which all generations are committed to live? After all, clean air protects Nature's regime of local precipitation, is part of the commons, and so is everyone's birthright.
4. Does the energy company have the people's permission—adults *and* children—to add pollution to the air when it then pollutes the soil that grows their food and affects the water everyone drinks (water needed for life itself, water for which there is no substitute) and thereby adds to the negative cumulative effects with which all

generations must live? As with air, healthy soil and adequate, clean water are part of the global commons through successive generations of children and so are everyone's birthright.

5. Does the energy company have the people's permission—adults *and* children—to add pollution to the stream, lake, bay, or Gulf of Mexico, thereby *irreversibly* adding to negative cumulative effects with which all generations are progressively obligated to live because the gulf and the ocean, of which it is a part, have no outlet and so concentrate not only the amount of toxic chemicals but also their increasing toxicity? I ask this because both the gulf and the ocean are part of the global commons and their biophysical health is everyone's birthright.
6. With the aforementioned biophysical constraints to environmental health and the people's birthright of the commons, who bears the burden of proof that no harm shall be done—those citizens who would protect the quality of their environment or *the parties who would pollute it?*[7]

With the above example in mind, we would be well advised to make the best possible use of what we know about assessing risks, provided we do *not* become enamored with an outcome as a "sure bet." Betting on the most knowledgeable probability of the outcome is all risk assessment is. The danger associated with the bet is our certainty of our *uncertain, ever-changing* knowledge—our ignorance of our ignorance, as it were. In the end, that ignorance too often has long-term negative effects with respect to the quality of our human habitat as well as that of all nonhuman creatures.

Summation

In Chapter 4, I described design as the organization of elements in some form of composition. Composition, in turn, is the combination of parts or elements to form the structure and function of a systemic whole, wherein the parts are individually and collectively constrained by their relationships to one another. In addition, whatever we introduce into the environment is immediately out of our control, as I demonstrated through the questions I asked in court.

Chapter 5, in turn, is a discourse on habitat; I point out that ultimately it is the constrained relationships of the parts that both creates and maintains the social-environmental sustainability of the system as a whole.

Notes

1. Chris Maser, Jon E. Rodiek, and Jack Ward Thomas. Cliffs, talus, and caves. In: *Wildlife Habitats in Managed Forest—The Blue Mountains of Oregon and Washington*, Jack Ward Thomas, technical ed., 96–103. USDA Forest Service. U.S. Government Printing Office, Washington, DC, 1979. Agricultural Handbook No. 553.
2. Chris Maser, Jack Ward Thomas, Ira David Luman, and Ralph Anderson. *Wildlife Habitats in Managed Rangelands—The Great Basin of Southeastern Oregon: Manmade Habitats*. U.S. Department of Agriculture, Forest Service, Pacific Northwest Forest and Range Experiment Station, Portland, OR. (1986). USDA Forest Service General Technical Report PNW-86.
3. Chris Maser. *Vision and Leadership in Sustainable Development*. Lewis Publishers, Boca Raton, FL (1998).
4. Chris Maser. *Ecological Diversity in Sustainable Development: The Vital and Forgotten Dimension*. Lewis Publishers, Boca Raton, FL (1999).
5. The foregoing comments on risk are gleaned from B. John Garrick. Society must come to terms with risk. *Corvallis Gazette-Times* (November 9, 1997).
6. Chris Maser. *Our Forest Legacy: Today's Decisions, Tomorrow's Consequences*. Maisonneuve Press, Washington, DC (2005).
7. The foregoing discussion of scientific and social questions is based on Chris Maser. *The Perpetual Consequences of Fear and Violence: Rethinking the Future*. Maisonneuve Press, Washington, DC (2004).

5

Habitat, the Language of Boundaries

Every ant knows the formula of its ant-hill,
every bee knows the formula of its beehive.
They know it in their own way, not in our way.
Only humankind does not know its formula.

Fyodor Dostoyevsky[1]

Personal boundaries are usually understood to be those lines of silent language that allow a person to communicate with others while simultaneously protecting the integrity of their personal space as well as the personal spaces of those with whom they interact. Beyond that, the language of boundaries transcends individual space to include familial space, cultural space, and even national space. Understanding personal boundaries among individuals of the same culture is hard enough; expanding that concept into a fluid ability to work among different cultures is more complex, and extending the concept still further to include the shared biophysical boundaries between Everyforest and Everycity is the most difficult of all to accomplish—especially for an industrialized people who are increasingly and rapidly isolating themselves from Nature.

These biophysical boundaries are encompassed in the general notion of *habitat* and need to be understood to further our exploration of the design interface between Everyforest and Everycity. Habitat consists of six things: (1) food, (2) water, (3) shelter, (4) space, (5) privacy, and (6) the connectivity that makes the previous five components readily available to an individual animal or human.

Food

Clearly, food is a major consideration when it comes to survival of any species, plant or animal, whether in a forest or city. But, there is a clear difference between food in a forest and that in a city. Namely, most animals in a forest must hunt for their food on a daily basis, although a few, such as some ants, actually grow their own, whereas others, such as the European red squirrel and beaver, store food in times of plenty for use in times of scarcity.

Everyforest

Animals within a forest consume a wide variety of foods. Some species change foods with the season—especially the females that continually choose the kind of vegetation with the highest nutritional content during the time they are pregnant and rearing their young. Other animals, such as the northern flying squirrel and long-footed potoroo, tend to be more restricted in diet, preferring truffles (the reproductive body of certain fungi) when in season and lichens when truffles are absent as a dining possibility.

Some animals, such as beaver, consume the inner bark of trees, whereas moles, shrews, and forest bats are largely insectivorous. Coyotes, golden jackals, raccoons, and bears eat a relatively wide variety of meats, fruits, and other plant materials. In contrast, weasels and members of the cat family are predominantly meat eaters. Some birds, such as the Cooper hawk and sparrow hawk, are primarily predators of small birds; others, such as the great-horned owl and eagle owl, eat whatever prey they can catch—from mice to gophers, rabbits, and domestic cats; from starlings to ducks and other owls. There are, in addition, otters that feed on fish, frogs, and the like, as well as animals, such as some mice and birds, that dine primarily on seeds.

Animals are adapted to feed in a variety of places: the air (birds and bats); in trees (squirrels and monkeys); on the surface of the ground (deer, elk, bear, and wild pigs); in snags (woodpeckers); under fallen trees (shrews); underground (moles); both underground and on the surface (gophers and lesser bamboo rats); and in the water (otters), to name a few possibilities.

Everycity

Many cities host a wide variety of culinary delights, from Chinese to American, Greek, Italian, Ethiopian, Japanese, Syrian, Indian, German, Vietnamese, Thai, and, of course, vegetarian. Although most people probably eat a variety of such foods, there are those who prefer a diet that is commensurate with their own culture, such as the kosher foods of the Jewish people.

Although most people in a city go to a variety of places to secure the food items they need to make a meal (grocery stores, homeless shelters, etc.), some grow their own in home gardens. Others spend most of their time going to their meals—eating out. When eating out, people sit in restaurants that are situated in the top floors of "skyscrapers," below the surface of the ground, next to and over water, next to railroad tracks, in bus stations and airports, on ferries, as well as fast-food joints wherever they happen to be, not to mention gobbling hot dogs in the bleachers at ball games.

Water

> Water is life, and the quality of water determines the quality of life.
>
> **Lake Superior Binational Forum, 2000.**

Water is an uncompromising necessity of life. Every living thing needs fresh water to survive.

Everyforest

In a forest, most vertebrate animals must *go to water*. In that sense, the proximity to water is paramount to which species can live where—wet forest, dry forest, in the water, next to water, at some distance from water. Clearly, every fish lives in water, although certain fish in the vicinity of the equator can live sealed in mud between wet seasons.

In the coniferous forests of western Oregon, the red tree mouse gets much of its moisture by licking rain and dew off the needles of the trees in which it lives, whereas the darkling beetle in the desert of Namibia in southern Africa stands on its head to collect the moisture in the breezes blowing off the Atlantic Ocean that condenses on its back and runs downward into its mouth. Moles living below ground get much of their moisture from the succulent food items they eat, such as earthworms. Other animals, say ravens and bats, fly to water. As well, bobcats, wolves, and deer travel some distance overland to find water.

Everycity

The difference between a forest and a city is that people have learned how to bring water to themselves, whereas animals in a forest are constrained in where they live by the presence or absence of water. Even though city dwellers are just as tied to the necessity of water for survival as forest dwellers, they need only go to a convenient location, such as a kitchen tap, turn a handle, and collect the water to drink in a glass or cup. Moreover, water is piped to the restaurant in the skyscraper or penthouse, just as it is to the abode below ground, the bathroom in the train stations, the coffee shop in the airport, and so on.

The Common Denominator between Everyforest and Everycity

Every living thing in the forest (plant and animal) is at the mercy of the weather when it comes to the availability of water. The same is true for people, although few know the ultimate source of water. If city folks were to think beyond the faucet, they might conceive the source of water to be the city's utility department, perhaps some reservoir or other, or maybe even a particular river or lake.

Be that as it may, the storage of a city's water originates in the soil of high mountains far from the tap you turn to fill a glass with this most precious of liquids. Water is stored in four ways: (1) in the form of snowpack above ground; (2) in the form of water penetrating deep into the soil, where it flows slowly below ground; (3) in belowground aquifers and lakes; and (4) in aboveground lakes and reservoirs.

Most water used in a city comes first in the form of snow, either at high elevations or northern latitudes, where it melts and subsequently feeds the streams and rivers that eventually reach distant communities and cities—rivers, such as the Columbia, Danube, Yangtze, and Ganges. Snowpack is aboveground storage that, under good conditions, can last as snowbanks late into the summer or even early autumn.

How much water the annual snowpack has and how long the snowpack lasts depends on seven things:

1. The timing, duration, and persistence of the snowfall in any given year.
2. How much snow accumulates during a given winter.
3. The moisture content of the snow. Wet snow holds more moisture than dry snow; moreover, snow disappears in two ways: sublimation and melting. *Sublimation* means that snow, accumulating in such places as the upper surfaces of tree limbs above the ground, evaporates and recrystallizes without melting into water and thus bypasses any role in a supply of available water.
4. Where the snowfall accumulates in relation to shade and cool temperatures in spring and summer, such as under the cover of trees and on north-facing slopes, versus open areas, where forests have been clear-cut, or south-facing slopes with no protective shade.
5. When the snow begins melting and the speed at which it melts. The later in the year it begins and the slower it melts, the longer into the summer its moisture is stored above and below ground.
6. How dry the soil is. A drought can dry the soil to the point that it will take some time (lag period) for enough water to infiltrate sufficiently to pass down the slope, thereby recharging streams and so becoming available to communities.
7. The health of the overall water catchment.

Although the first points seem self-evident, the last one requires some explanation. In dealing with the health of water catchments, one needs to account for those in both high and low elevations.

High-Elevation Water Catchments

How we treat high-elevation forests (and those at more northerly latitudes) is how we treat a major portion of the most important sources of our supply of potentially available water, which originates as snow.

With the advent of late spring and early summer, the snow begins to melt and gradually infiltrates the soil until every minute nook and cranny is filled to capacity with the precious liquid, which all the while obeys the unrelenting dictates of gravity as it journeys along ancient geological paths toward the streams and rivers of the land on its way to the sea from whence it came. As gravity pulls the water downward through the soil, the slowly melting snow continually fills the void left by the departing liquid. In this way, the melting snows of winter feed the streams of late summer and autumn, thereby bringing water to human communities.

Because logging roads progressively fragmented the once contiguous forest and clear-cut after clear-cut merged into gigantic, naked mountain slopes, the snow melts earlier and faster than it once did. Now, the water-holding capacity of the soil is often reached in late May and early June. The inability of the soil to absorb these great pulses of water causes most of it to flow over the surface of the ground, where it rushes down streams and rivers, speedily fills the reservoirs to overflowing, and so is lost to the streams and forest animals, as well as human communities, when they need it most—in late summer and autumn.

To help you visualize the downhill journey of water, imagine a large, rotten log with both ends cleanly cut off, lying up and down on a steep slope under the canopy of an ancient forest. If the snow is deep enough, the melting water infiltrates the log at its upper end and is gradually pulled downward through the interior of the log by the inexorable pull of gravity until it drips out the bottom of the cut face at the log's lower end.

There is, however, a caveat to this phenomenon. If the snow is deep enough to cover the upper end of the log, the log can absorb the same amount of water that drips out the bottom just as long as the supply lasts. But, as soon as the snow is gone, the available supply of water is cut off, and that remaining in the log will eventually drip out the bottom end without being replenished. Therefore, the longer the snow lasts at the upper end of the log, the longer the log can act as a conduit for the water infiltrating its upper end, passing through its length, to drip out its lower end. Conversely, the faster the snow disappears from the log's upper end, the faster the supply of water is cut off, the quicker the log progressively dries out, even as water continues to drip out the lower end. That also will cease because there is no replenishment for the limited supply of water pulled through the log by gravity.

Similarly, the downspout from a roof will keep water flowing just as long as rain falls on the roof or snow accumulates and is there to melt. The concept is the same—supply and availability.

Thus, when contemplating the supply of water for a city, humility, wisdom, and long-term economics would dictate that some forested areas, particularly at high elevations, are left intact—not cut even once, despite the perceived, immediate, short-term dollar value of the wood fiber. To protect such areas for the accumulation and storage of snow, hence water, will require a drastic shift in thinking because, at present, the only economic value seen in high-elevation forests is the immediate extraction of wood fiber. Nevertheless, the logging roads in and clear-cutting of high-elevation forests that catch and store water affects all human communities, from the smallest rural town to the largest city.

Low-Elevation Water Catchments

Most low-elevation water catchments, which may or may not be forested, need to be much larger in area than a high-elevation catchment in order to collect and store the same amount of water. Although snow may not be as important for the storage of water in low elevations, the ability of water to infiltrate deep into the soil is equally important. The storage of water at low-elevation, nonforested areas is often in wetlands, subterranean aquifers, and lakes, as well as in aboveground reservoirs. Regardless of where the water catchment is located, roads and urban sprawl have a tremendous negative effect on both the quality and the quantity of water that ultimately reaches a community or city.

Shelter

Shelter comes in many forms, in both Everyforest and Everycity. Yet, the similarities are seldom the point of focus.

Everyforest

Animals in a forest live in nests in the tops of live trees (squirrels), whereas others use the broken tops of tall trees to support their nesting platforms (osprey, eagles, and owls). Still other animals live in holes in snags (woodpeckers, squirrels, weasels, and bats) and under loose bark attached to snags (bats and, in some cases, treecreepers—a little bird). In fact, female northern flying squirrels occupy cavities made by the pileated woodpecker to the exclusion of males because the well-insulated cavities are superb for raising their young. With the eviction rule in force, male squirrels have little choice but to accept cavities of lesser quality, where two, three, four, or more huddle together for warmth.[2]

Some birds, namely swifts, nest inside large, hollow snags; likewise, some bats form nursery colonies within such snags. Several birds or mammals, often of different species, occasionally use a particularly large snag of good quality simultaneously. Still other animals live in cavities in the trunks of solid snags (woodpeckers and tree swallows) or hollows in the bases of live trees, where the trunk joins the large roots (hibernating bears, resting marten or fisher).

Animals live in holes in the ground and build tunnels from one belowground chamber to another (gophers with their separate nesting chambers, food storage chambers, and toilet chambers). Some burrowers have tunnels large enough to accommodate other animals; tunnels of the mountain beaver (a terrestrial rodent) are used by mice, spotted skunks, snowshoe hares, brush rabbits, weasels, and so on.

Animals also live in caves (bats and bushy-tailed woodrats), in crevices in the faces of cliffs (mice, bats, and bushy-tailed woodrats), and on ledges protruding from cliff faces (ravens and golden eagles). Some are even colonial in that they build their nests together (cliff swallows that attach their mud nests to the undersides of ledges on cliff faces and sand martins that make their nests in neighboring burrows in the banks of rivers). One mammal (the Himalayan pika, a small "rock rabbit") lives within the areas of broken, jumbled rock, called *talus*, at the base of cliffs.

Everycity

Although you may not think so, people in a city emulate virtually every major shelter found in a forest. People live in well-insulated buildings constructed of wood (tree), stone or concrete (cave), skyscraper (hollow snag), and so on. They live from the ground floor (the cavity at the base of a tree) to the penthouse (the top of a tree). They live in colonies (apartments), in burrows (basement abodes), and even construct tunnels (hallways) that connect one chamber to another (belowground restaurants, public toilets, shops, subways, and so on).

The Common Denominator between Everyforest and Everycity

Shelter serves two primary purposes: protection from the weather and a purveyor of privacy. With respect to protection from the weather, animals have naked skin (amphibians), scales (reptiles), feathers (birds), or hair (mammals). Amphibians and reptiles are cold blooded inasmuch as they control their bodily temperatures through the ambient temperature of their surroundings. Birds and mammals carry their weather-ameliorating "clothing" with them in feathers and hair. People in a city, on the other hand, either make or purchase clothing with which to keep comfortable in various kinds of weather; in that sense, people carry their artificial, microweather with them as they go about their daily routines.

Despite feather, fur, and clothing, shelter is needed to protect both forest animals and city people from the cold of winter and the heat of summer. This type of shelter is termed *thermal cover* (as opposed to *hiding cover*) and is provided by thickets of trees with dense crowns. Topographically sheltered areas also provide thermal cover.

To illustrate, thermal cover for deer and elk is provided in summer by thickets of trees and other dense vegetation, where they can lie down and conduct the heat of their bodies into the cool of the soil. In winter, such vegetation protects them from the chilling winds and thereby conserves their body heat. In both cases, there is a required proximity of thermal cover and foraging areas, which for us humans would be like requiring a clothing store immediately adjacent to a restaurant because we rely on clothing first and foremost to regulate our body temperatures. But when that fails, we resort to warm buildings in winter and cool buildings in summer.

On the other hand, gophers regulate the temperature of their abodes by opening and closing the various entrances into their tunnel system, an act similar to a homeowner adjusting the thermostat to control the temperature. When sufficient shelter is not available, both animals and people die from exposure to either excessive heat or penetrating cold.

Space

Space is usually taken for granted when discussing habitat, but with the continual fragmentation of landscapes, habitable space is becoming a limiting factor for an increasing number of species as people progressively alter the overall environment to suit their desires and perceived necessities based on those desires.

Everyforest

Space in a forest relates directly to the adaptability (versatility) a particular species exhibits with respect to the number of plant communities and developmental stages within a plant community that it can use for feeding and reproduction. The greater the number of plant communities and stages a species can use, the more adaptable it is. Conversely, the fewer the communities and stages a species can use, the less adaptable it is.

Space is further divided into that portion of the habitat used for feeding and that portion used for reproduction. In general, a species is most adaptable in terms of its habitat for feeding and least adaptable in terms of its habitat for reproduction. Put differently, a species can normally use a *greater* number of plant communities and developmental stages within a plant community for feeding than it can for reproduction. For example, the common flicker (a woodpecker) in northeastern Oregon can use twelve plant communities and developmental stages for feeding and nine for reproduction, which, when added together, gives the flicker an adaptability score of *twenty-one*.

In contrast, the white-headed woodpecker in the same region can use four plant communities and developmental stages for feed and four for reproduction, giving the white-headed woodpecker an adaptability score of *eight*.[3]

Beyond that, a particular structure within the developmental stage of a plant community may be a limiting factor with respect to the ability of a species to reproduce. A case in point is a large pileated woodpecker of North America or green woodpecker of Eurasia, both of which require a snag that is twenty-inches in diameter for nesting (such as found in an old forest) and simply cannot fit into a snag that is eleven inches in diameter (such as found in a stand of young trees). A downy woodpecker, on the other hand, can nest in a snag that is eight inches in diameter but can also fit quite nicely into a snag that is twenty-inches in diameter. Clearly, a downy woodpecker (which can nest in both a young forest and an old forest) is more adaptable reproductively than is the pileated woodpecker.

In turn, the notion of a habitat for feeding and a habitat for reproduction corresponds well to an animal's home range and territory. A *home range* is that area of an animal's habitat in which it ranges freely throughout the course of its normal activity and in which it is free to mingle with others of its own kind—the shared use of an area. A *territory*, in contrast, is that part of an animal's home range that it defends, for whatever reason, against others of its own kind—analogous to private property. This defensive behavior is most exaggerated and noticeable during an animal's breeding season.

Everycity

How might this notion of "space" translate into human terms within a city? How does it relate to human adaptability (versatility)? Think about the average home. There is a room, or maybe two, dedicated to social interaction: living room and family room. The kitchen and dining room are used for the preparation of food and eating. One or more bathrooms are used to maintain cleanliness and eliminate bodily wastes. And, the most private area of a house is the bedroom, which is used for sleeping and reproduction.

As for home range and territory, suppose it is Saturday morning, and you leave your home to take care of a few errands. You simply go about your business without paying much attention to what is going on around you or to the people you pass unless you happen to meet someone you know. In general, you are simply engrossed in what you are doing. When you have finished your errands, you start home.

The closer you get to your neighborhood, the more alert you unconsciously become to changes around you, such as the new people moving in two blocks away. This "protective feeling" becomes even more acute as you approach the area of your own home and notice a car with an out-of-state license plate parked in your neighbor's driveway. You get out of your car and immediately notice, perhaps with some irritation, that someone's dog has visited your lawn while you were gone. If the dog had anointed someone else's yard with its leavings, you probably would have paid less attention or none at all.

The same general pattern extends into your home. How well you know someone and how comfortable you feel around them determines the freedom with which they may interact with you and your family as well as how free they are to move about in your house. On the other hand, you are most particular—and protective—about your ultimate private spaces—your bedroom and your physical being.

Privacy

Privacy is an often-overlooked component of habitat. Although it is well known that people cherish privacy (some more than others), it is less well known that animals also need privacy—hence the ubiquitous nature of territorial behavior, especially as it relates to reproduction.

Everyforest

In addition to the aforementioned thermal cover, animals require "hiding cover." Hiding cover can be vegetation, topography, a burrow, a nest, a crevice in the face of a cliff, a cavity in a tree, a secluded space within a slope of fractured rock (talus) at the base of a cliff—in essence any structure, usually darkened, that allows an animal to choose whether it wants to be seen or otherwise disturbed. This kind of cover both hides an animal from predators and provides privacy for resting, reproducing, and rearing young.

Everycity

People have the same psychological need for privacy that other animals do. And, they have a variety of means for creating such a sanctuary; drawing curtains closed or closing Venetian blinds or other window coverings is a common method of excluding the "outside world" and creating a private space enclosed within one's home. Moreover, within a house there generally is the privacy of one's own room, although such may not be the case in a dysfunctional family, where boundaries often have no meaning.

Another way people in large cities find privacy is to withdraw into themselves. I observed this kind of behavior numerous times while riding the commuter trains in Tokyo, Japan. A person would enter a train, select a seat if one was available or stand if one was not. They then closed their eyes and remained quiet—shuttered as it were against the outer world—until it was time for them to leave the train. I have no idea how they knew when the station arrived at which they wanted to depart, but they always seemed to know. In a strange way, I found this behavior to be a most profound kind of privacy.

Connectivity

Connectivity among the various habitat components is critical to the overall quality of a habitat in our social-environmental landscapes. The greater the connectivity among habitat components, the greater is a habitat's quality and the more continuous its use will be. The extent and quality of the connectivity among habitat components (such as water, food, shelter, space, and privacy) determines which person or species can live where.

Connectivity in Everyforest

One circumstance—the arrangement of habitat components across the landscape—is vastly different between Everyforest and Everycity in that people rearrange land on which to design, build, and place their own shelters as well as bring food and water into them from afar. Moreover, people can routinely store excess food for long periods in freezers, regardless of weather or climatic conditions, but most animals must find their required ration on a daily basis. In contraposition, Nature creates habitable areas within a forest (a snag of the right size, a fallen tree under which to dig, etc.), not the animals that need them. Animals in a forest must also go to the food and water, and many must find existing shelter as well.

Another circumstance—the connectivity of habitat components in and across the landscape—is equal in importance. Proximity to food, water, and shelter is critically important to the most sedentary and highly adapted species (such as salamanders). Proximity to food, water, and shelter is progressively less important as one proceeds toward the most wide-ranging and adaptable species (such as jackals, jaguars, and deer) that can travel great distances in short periods of time. Nevertheless, there is a condition imposed on the latter. Wide-ranging species that live in ever-more-fragmented habitats require safe corridors of travel through "hostile" terrain from one habitat component to another. With this in mind, let us take a quick look at a few of the possibilities.

Water

A salamander or a frog cannot travel a mile to get a drink of water when it needs it, but a deer, bear, or lion can. By the same token, an eagle can fly a great distance to water, but most forest bats require drinking water within roughly a quarter of a mile from their daytime roosts or nursery colonies.

As an aside, an adaptation to the problem of getting water is that of the sand grouse (a bird the size of a small chicken) I saw in the Egyptian and Nubian deserts. Sand grouse nest well away from water, minimizing the potential predation on their young, which require water nonetheless. Sand grouse therefore fly great distances to water, soak their breast feathers in the precious liquid, and fly back to their nests to give their offspring a drink of the water stored in their feathers.

Food

Grazers, such as deer and some antelope, require food and cover in close proximity because they are vulnerable to predators while feeding. Ergo, they venture from cover, eat rapidly, and return to cover, where they chew their cud (regurgitated food), an act that corresponds to eating their food a second time, but in relative safety. Deer also migrate from a summer range to a winter range and back again, meaning they need a good-quality corridor of habitat within which to travel.

Small forest animals, such as amphibians, shrews, and mice, use the open, downhill side of logs as a protective cover while navigating the surface of the forest floor in search of food—an idea emulated by the concrete tunnels open on the downhill side that protect traffic from snowdrifts and avalanches in the Austrian Alps and higher elevations of Japan. Ants, on the other hand, create open "highways" in areas of dense, herbaceous vegetation, both in the first stage of development in a forest and in closed-canopy tropical forests; these highways are often crowded with hundreds of individuals going far afield from their colony in search of food. In addition, some rodents make aboveground runways from one fallen tree to another through the herbaceous ground cover but construct belowground burrows at other times, thereby connecting one place with another. Belowground burrow systems are also the preferred mode of travel for the pocket gophers as they forage, although they construct aboveground burrow systems through the snow in winter. These snow tunnels are somewhat analogous to the heated, overhead "tunnels" through which people can comfortably cross over busy streets in Duluth, Minnesota, even in winter, when the wind chill factor is dangerously bone chilling.

Other animals, like otters, seem to have a wanderlust and travel great distances to fulfill their life's requirement, which includes a prodigious amount of "playtime." The marten, cousin of the otter, is also a traveler but stays within forested areas, where its prey base is located, and is thus constrained in its movements by such habitat alterations as clear-cut logging because the marten will not readily venture into such open areas. Jungle cats, which are stalkers, require sufficient cover in order to ambush their prey, but wolves, which are chasers, require open areas to "run their prey to ground."

Shelter

Some small mammals have everything neatly packaged. Water voles live along streams in the higher elevations, where they use both the waterway and runways through herbaceous vegetation of the stream's banks to move about and obtain their vegetarian food. In addition, they burrow into the stream banks, where they build snug nests—having all their life's necessities in immediate proximity.

Birds and bats, on the other hand, spend time away from their living quarters (cavities in snags, nests in shrubs and trees, caves, and so on) while they forage and quench their thirst. Being aerially mobile, they are relatively free to move about and stitch their habitat requirements together with greater facility than some of their earth-bound kin.

Still others, such as snow leopards and elk, traverse great distances in search of food, water, and shelter. To them, "home is where their rump rests" on any given day.

Space

Not only is the arrangement of habitat components important to the animals in a forest but also important is the habitat's extent in area. To illustrate, the aforementioned water vole requires but a small area along a mountain stream in order to have a viable lifestyle. Compared to the water vole, coyotes are exceedingly adaptable, relatively wide-ranging, independent animals that can seemingly survive anywhere, including in the suburbs of Los Angeles, California, where they help themselves to garden produce, scraps of human food, as well as that left out for pets. In addition, they are quite willing to eat the neighborhood cats, chickens, ducks, or any other handy foodstuffs.

Roaming the country singly, in pairs, or as family groups, coyotes prey on a wide array of kinds and sizes of animals. Whereas pups prey on grasshoppers, adults tackle animals as large as mature mule deer and yearling domestic cattle, or any other seasonal morsel they deem tasty. Coyotes are also adept at dining on fruits, a habit that earned them the nickname "melon wolves" in some parts of their geographical distribution because they steal from farmers' fields.

As a generalist, the coyote can survive under a wide range of environmental conditions, from Texas to Alaska and from the Pacific Coast to the eastern seaboard. Their arrival in Alaska and on the eastern seaboard within relatively recent years is due primarily to the clear-cutting of vast areas of dense forest. Our social activities have opened thousands of square miles for the coyote to inhabit because of its extraordinary adaptability. And, due to its wide array of possible food items, the coyote can make use of a broad variety of energy sources.

Wolves, on the other hand, are social animals that live in packs. Compared with coyotes, their group life places limits on their ability to use habitats. This means a far greater number of coyotes than wolves can live in a given location because a pack of wolves acts as a single, large organism and so requires a much vaster area for hunting. A pair of coyotes can live on rabbits and fruits in season, but a pack of five to seven wolves, each of which is much larger than even a big coyote, acts collectively as a single, large animal that must continually secure prey the size of deer, elk, barren-ground caribou, and moose.

A pack of wolves has a much harder time staying fed than does a pair of coyotes. It takes far more time, energy, and trials for a pack of wolves to select, chase, and bring down large prey at any time of the year than it does for a coyote, which at certain times of the year can make do quite nicely on a diet of grasshoppers, meadow mice, and ripe berries.

As a specialist, the wolf is fitted to a narrow set of environmental circumstances and can survive only if it finds sufficient prey large enough to feed the pack as a whole. Because of this, the wolf has a limited range of prey items to which it is effectively adapted as sources of energy, and it can neither fit itself to a wide variety of conditions nor can it fit a wide variety of conditions to itself.

The wolf, as a highly *adapted, social specialist*, requires a large area (compared to a pair of coyotes) to keep the pack fed. This requirement makes the wolf vulnerable to extinction due to humanity's continual encroachment on and fragmentation of its habitat, while the coyote, as a supremely *adaptable, individualistic generalist*, is likely to outsurvive humanity itself. This is particularly evident as the wolf's geographical range shrinks in the face of societal pressures on the landscape, while the coyote's geographical range increases in concert with those same pressures.

Privacy

With respect to privacy, the little water vole probably has the most easily satisfied requirement in that privacy is only a burrow away. The medium-sized coyote, on the other hand, is so adaptable and individualistic that it can find privacy almost anywhere. For a large pack of wolves in a progressively fragmented habitat, privacy is becoming increasingly rare and is part of the reason so much of their former geographical range is no longer suited for them to inhabit.

Connectivity in Everycity

Within a city, water is brought into private homes, public buildings, and even city parks. Food, including water and other liquid refreshments, is readily available in stores and restaurants, as well as from sidewalk vendors. Shelter comes in many forms in a city: buildings for living (single-family homes, duplexes, apartments, etc.), buildings for working (offices), buildings for eating (restaurants), buildings for learning (schools) buildings for the ill (hospitals), buildings for the mentally ill (mental institutions), buildings for entertainment (theaters), buildings for criminals (jails and prisons), buildings for exercise (gyms), buildings for worship (churches), as well as buildings for transportation (bus stations, train stations, airports—and here one might include subway tunnels).

With respect to space, how the various kinds of buildings are distributed across space determines how accessible the habitat components are to an individual. For example, water, food, and shelter arranged for the purpose of multiple-use in a relatively small area, say walking distance, as in many European villages, allows the inhabitants maximum flexibility in meeting their daily living requirements with a minimum of effort—a well-connected habitat. On the other hand, typical urban sprawl in the United States means extended trips from one necessary habitat component (home) to another (grocery store, doctor's office, school, etc.), often through "hostile" terrain, where a person is constrained to roads, rails, or subways, along which no available food, water, or shelter exists that a person could access should the need arise—a highly fragmented habitat, often between home and work.

The foregoing scenario is forcing village people (analogous to more sedentary species that meet their habitat requirements in a relatively small space, such as a salamander, frog, mouse, rabbit, or naked mole rat) to become far more wide ranging, like wolves, wildebeest, wolverines, or tigers. Although people have a choice during the planning phase of developments, once the results are ensconced on the landscape, the individuals who live with the consequences have little or no choice.

With respect to space, the kind of shelter one occupies determines how much or little privacy one has. People in a single-family home have the most privacy. A person without means has relatively less privacy, whereas someone in a maximum-security prison has the least.

Of Everyforest and Everycity

Although forest-dwelling animals and city-dwelling people have the same array of habitat requirements, they are fulfilled differently. People not only have far greater control over their environment and so their collective habitat requirements than do animals but also have control over the habitat requirements of the animals. Consider that clear-cutting a forest is like razing a centuries-old, mixed-use portion of a town. If the clear-cut is then planted to a single-species, economically driven monoculture, that is the same as rezoning the area from "mixed-use residential" to "commercial" and immediately constructing a shopping mall.

In both cases, the residents have been involuntarily displaced, and some may even become permanently homeless. This vulnerability is exactly the circumstance faced by most forest-dwelling animals when their mixed-use habitat (the old forest) is cut out from under them and replaced by a sterile "fiber farm" that is subjected to an ongoing forty- to sixty-year, cut-plant-cut-plant cycle. This repetitive cycle is somewhat analogous to the built-in obsolescence of today's urban environment.

Moreover, an old forest differs significantly from a young forest in species composition, structure, and function. Most of the obvious differences can be related to four structural components of the old forest: large live trees, large snags (standing dead trees), large fallen trees on land, and large fallen trees in streams.

On land, this large, dead woody material is a critical carryover component from old forests into young forests. When snags are removed from a short-rotation stand following liquidation of the preceding old growth (coniferous forest, in this case), 10 percent of the wildlife species (*excluding* birds and fish) will be eliminated; 29 percent will be eliminated when both snags and large fallen trees (logs) are removed from the intensively "managed" fiber farm.[4]

The connectivity—accessibility—of habitat components is particularly important for the resident population, regardless of species, because each habitat has a biological carrying capacity, meaning a finite number of individuals that can live in a particular area without altering it to their detriment. If, however, you, the reader, either grew up in a city or now live in one, you may be familiar with habitat corridors in urban settings.

Birds in urban landscapes generally occupy parks, which are analogous to forest fragments, whereas tree-lined streets form linear corridors that connect the fragments within the urban matrix. To understand the species-habitat dynamics of an urban setting, a study conducted in Madrid, Spain, examined the effects of street location within the urbanscape, vegetative structure along the streets, and human disturbance (pedestrian and automotive) on bird species richness within the street corridors. In addition, the birds' temporal persistence, density of feeding and nesting guilds, and the probability of a street's being occupied by a single species were also taken into account.

The number of species increased from the least-suitable habitats (streets without vegetation) to the most suitable habitats (urban parks), with tree-lined streets being an intermediate landscape element. Tree-lined streets that connected urban parks positively influenced the number of species within the streets' vegetation, as well as species persistence, population density, and the probability that the individual species would continue to occupy the streets. That said, human disturbance did exert a negative influence on the same variables.

Wooded streets could potentially function as corridors that would allow certain species to fare well by supporting alternative habitats for feeding and nesting, particularly those birds that feed on the ground and nest in trees or tree cavities. Local improvements in quality and complexity of the vegetation associated with certain streets, as well as a reduction in the disturbance caused by people, could exert a positive influence on the regional connectivity of streets as a system of urban corridors for birds. Because of the differential use of corridors by species with various habitat requirements, streets as habitat corridors could be further improved by taking the requirements of different species into account.[5] Here, it is instructive to observe communities of birds.

First, there is the resident community or that group of birds inhabiting the area to which they have a strong sense of year-round fidelity. In order to stay throughout the year, year after year, they must be able to meet all of their ongoing requirements for food, water, shelter, space, and privacy. These requirements become most acutely focused during the time of nesting, when young are reared, and during harsh winter weather.

Then, there are the summer visitors that overwinter in the southern latitudes and fly north to reproduce. They arrive in time to build their nests, rear their young, and in so doing must fit in with the year-long residents without competing severely for food, water, shelter, space, or privacy—especially during nesting. If competition is too severe, the resident community will decline and perhaps perish through overexploitation of the habitat by summer visitors, which have no lasting commitment to a particular habitat.

There are also winter visitors that spend the summer in northern latitudes, where they rear their young, and fly south in the autumn to overwinter in the same area as the yearlong residents, but after the summer visitors have left. They must also fit in with the yearlong residents without severely competing for food, water, shelter, space, and privacy during times of harsh weather and periodic scarcities of food. Here, too, the resident community will decline and perhaps perish if overexploitation of the habitat through competition is too severe. And, like the summer visitors, the winter visitors are not committed to a particular habitat but use the best of two different habitats (summer and winter).

On top of all this are the migrants that come through in spring and autumn on their way to and from their summer nesting grounds and winter feeding grounds. They pause just long enough to rest and replenish their dwindling reserves of body fat by using local food, water, shelter, and space, to which they have only a passing fidelity necessary to sustain them on their long journey.

The crux of the issue is the carrying capacity of the habitat for the yearlong resident community. If food, water, shelter, space, and privacy are sufficient to accommodate the yearlong resident community as well as the seasonal visitors and migrants, then all is well. If not, then, in effect, each bird in addition to the yearlong residents causes the area of land and its resources to shrink per resident bird. This in turn stimulates competition that, under circumstances of plenty, would not exist. But, if such competition causes the habitat to be overused and decline in quality, the ones who suffer the most are the yearlong residents for whom the habitat is their sole means of livelihood.

With respect to cities, the same problem is reflected in a shift from a longtime, resident population to a predominantly transient one. The notion of transient can be thought of in two ways: short-term occupancy of an area and long-term occupancy of an area.

In the former, people simply move into an area to exploit opportunities or avail themselves temporarily of existing services. By long-term transience, I mean the people who are quite willing to live in a community while earning a living, but the community does not hold sufficient value for them to retire there. So, while they take advantage of the town as a place to work, they plan to move somewhere else upon retirement. Here, the result can be disastrous to the resident community because it is based on permanent *taking* and minimal, temporary *giving*.

Many people today think of competition when a discussion of habitat arises, both in terms of animals and people. The notion of competition seems pervasive, in part because of humanity's massive habitat alterations that often pit heretofore noncompetitive species (or, in the case of people, individuals) against one another for shrinking resources. There are alternatives, despite such competition, one of which is "resource partitioning," meaning different species are variously adapted to exploit Nature's bounty without direct competition. It works as follows:

If two species are active at the same time (e.g., the tree squirrel and a ground squirrel), they use the habitat differently. The ground squirrel nests and reproduces in burrows and has a more catholic diet than the tree squirrel. The tree squirrel in turn nests and reproduces in the treetops. In addition, the ground squirrel hibernates during the winter when food is scarce, whereas the tree squirrel stores food for winter and so is active all year.

In contrast, if two species use a habitat in a similar manner, they are separated by their time of use. To illustrate, a tree squirrel and a flying squirrel both nest and reproduce in the treetops and have a propensity for truffles (the belowground fruiting bodies of certain fungi). They do not, however, compete directly for the same food because the tree squirrel is active during the day and the flying squirrel is active at night.

When it comes to garnering the economic means to live in a city, people also partition resources. Say, for instance, that a family of twelve all want to earn a living in the same city. How might they go about doing it? The father could be a dentist, the mother a real-estate broker, one son is a banker, the second a lawyer, the third a policeman, the fourth a mechanic, the fifth a preacher; one daughter is a druggist, the second a teacher, the third the mayor, the fourth a nutritionist, and the fifth a postal clerk. In this way, the entire family can pursue their respective livelihoods, complementing one another's professional goals by working in different buildings, in different parts of town, doing different things, instead of competing with one another for limited jobs. This scenario is similar to that of the ground squirrel and the tree squirrel.

But, what would happen if three of the siblings wanted to be doctors in the same hospital? They could easily accomplish this by working different shifts: day, swing, and graveyard—a condition similar to the diurnal tree squirrel and the nocturnal flying squirrel.

Summation

In this chapter, I discussed the concept of "habitat" by describing each of its components so you can see what they are, how they function, and why each is critically important to the whole. As it turns out, the arrangement of habitat components across the landscape is vastly different between Everyforest and Everycity in that Nature creates habitable areas within a forest (a snag of the right size, a fallen tree under which to dig, etc.), not the animals required to use them. This means forest-dwelling animals are constrained both by where they can find suitable areas for shelter and by where they can find available food and water. In contraposition, people rearrange land on which to design, build, and place their own shelters, as well as bring their food and water into them from afar.

Another circumstance, however, is equally important, namely, the connectivity of habitat components in and across the landscape. Proximity of food, water, and shelter is critical to the most sedentary, highly adapted species (such as salamanders), which is analogous to residents in a mixed-use, residential part of town—a well-connected habitat. While the proximity of food, water, and shelter is progressively less important as one proceeds toward the most wide-ranging and adaptable species (such as coyotes and, increasingly, people) that can travel great distances in short periods of time, nevertheless, the continual fragmentation of habitats is one of the world's most serious problems. Fragmented habitats increasingly require species to have safe corridors of travel through "hostile" terrain among habitat components: food ↔ water ↔ shelter ↔ food.

In short, habitat forms that part of the social-environmental landscape where we all live. Omit a single component of habitat (e.g., food, water, or shelter; all coupled with money in a city) and life cannot exist—either on the individual level or, if the omission is drastic enough, on the species level. Such connectivity of habitat components is determined by patterns across the landscape, those created by Nature and those designed by humans; these patterns are the subject of Chapter 6.

Notes

1. Fyodor Dostoyevsky. Dostoyevsky: Humankind and Its Formula. http://donaldkim.wordpress.com/2008/12/11/dostoyevsky-humankind-and-its-formula/ (accessed on January 19, 2009).
2. Chris Maser, Ralph Anderson, and Evelyn Bull. Aggregation and sex segregation in northern flying squirrels in northeastern Oregon, an observation. *Murrelet*, 62 (1981):54–55.

3. Jack Ward Thomas, Rodney J. Miller, Chris Maser, and others. Plant communities and successional stages. In: *Wildlife Habitats in Managed Forest—The Blue Mountains of Oregon and Washington*, Jack Ward Thomas (technical ed.), pp. 22–39. USDA Forest Service. U.S. Government Printing Office, Washington, DC, 1979. Agricultural Handbook No. 553.
4. Larry D. Harris and Chris Maser. Animal community characteristics. In: *The Fragmented Forest*, Larry D. Harris (ed.), pp. 44–68. University of Chicago Press, Chicago, 1984.
5. The discussion of the Madrid, Spain, study is based on Esteban Fernández-Juricic. Avifaunal use of wooded streets in an urban landscape. *Conservation Biology*, 14 (2000):513–521.

6

Patterns across the Landscape

> To the poet, to the philosopher, to the saint, all things are friendly and sacred, all events profitable, all days holy, all people divine. For the eye is fastened on the life, and slights the circumstance. Every chemical substance, every plant, every animal in its growth teaches the unity of cause, the variety of appearance.
>
> **Ralph Waldo Emerson**[1]

The spatial patterns on landscapes result from complex interactions among physical, biological, and social forces. Most landscapes have been influenced by the cultural patterns created by human use, such as farm fields intermixed with the patches of forest that surround a small town or large city. The resulting landscape is an ever-changing mosaic of unaltered and manipulated patches of habitat that vary in sizes, shapes, and arrangements.

Our responsibility now is to make decisions about patterns across the landscape while accounting for the consequences of our decisions on the land's productive capacity for generations to come. Although the decisions are up to us, one thing is clear: While the current trend toward homogenizing the landscape may help maximize short-term monetary profits, it devastates the long-term biological sustainability and adaptability of the land and consequently diminishes the land's long-term productive capacity.

Relationship and Pattern

Stability flows from the patterns of biophysical relationships that have evolved among the various species or developed among people. A seemingly stable, culturally oriented system—even a very diverse one—that fails to support these ecologically coevolved (or personally developed) relationships has little chance of being sustainable, especially when a long-term timescale is taken into account.

What is the relevance of time in creating stable relationships? How might this work in a larger collection of people, say a community? Community, as English historian Arnold Toynbee said of civilization, "is a movement and not a condition, a voyage and not a harbor." But "community" is a deliberately different word than "civilization" or even "society." Although community may refer to neighborhoods or workplaces, to be meaningful it must imply membership in a human-scale collective, where people encounter one another face to face.

Community in a politically free society is a group of people with similar interests living under and exerting some influence over the same government in a shared locality. Because they have a common attachment to their place of residence, where they have some degree of autonomy, they are the resident community.

People in such a community share social interactions with one another and organizations beyond government, and through such participation are able to satisfy the full range of their daily requirements within the local area. The community also interacts with the larger society, both in creating change and in responding to it. Finally, the community as a whole interacts with the local environment, molding the landscape within which it rests and is in turn molded by it. In this sense, a sustainable community is about the oneness of the whole and the wholeness of the one.

A community also has a history that has to pass from one generation to the next if the community is to know itself throughout the passage of time. Because history is a reflection of how we see ourselves, it goes to the very root by which we give value to things. Our vision of the past is shaped by, and in turn shapes, our understanding of the present—those images we carry in our heads by which we decide what is true and what is not.

When the continuity of a community's history is disrupted, the community suffers extinction, or at least a blurring of identity, and often begins to view its landscape as a commodity to be exploited for immediate financial gain rather than as an inseparable extension of itself. When this happens, community is destroyed from within because trust is withdrawn in the face of growing economic competition.

It seems clear, therefore, that the concrete nature of a community literally cannot extend beyond its local place and history. Although the notion of community can be expanded, it is only metaphorical because, as author Wendell Berry said, "The idea of a national or global community is meaningless apart from the realization of local communities."

For a true community to be founded in the first place and to be healthy and sustainable, it must rest on the bedrock of mutual trust among its members, as eloquently penned by Wendell Berry:

> A community does not come together by covenant, by a conscientious granting of trust. It exists by proximity, by neighborhood; it knows face to face, and it trusts as it knows. It learns, in the course of time and experience, what and who can be trusted. It knows that some of its members are untrustworthy, and it can be tolerant, because to know in this matter is to be safe. A community member can be trusted to be untrustworthy

> and so can be included. But if a community withholds trust, it withholds membership. If it cannot trust, it cannot exist.[2]

In sum, community is relationship, and meaningful relationship is the foundation of a healthy, sustainable community. In this connection, Ralph Waldo Emerson felt that, "It is one of the most beautiful compensations of this life that no man can sincerely try to help another without helping himself."[3] And William James was of the opinion that "wherever you are, it is your own friends who make your world."[4]

Community also reminds us that the scale of effective organization and action has always been small local groups. As anthropologist Margaret Mead noted: "Never doubt that a small group of thoughtful, committed citizens can change the world; indeed it is the only thing that ever has."[5]

When a community loses the cohesive glue of trust embedded in its fundamental values, it loses its identity and is set adrift on the ever-increasing sea of visionless competition both within and without, where "growth or die" becomes the economic motto driving the cultural system. Such blindness inevitably rings the death knell of community.

To create viable, culturally oriented landscapes (such as a city and its surroundings), requires us, beginning now, to work toward a connectivity of habitats because ecological sustainability and adaptability depend on the integration of the landscape. To achieve this kind of landscape, we must ground our culturally designed landscapes within Nature's evolved patterns if we are to have any chance of creating a quality environment that is both pleasing to our cultural senses and ecologically adaptable.

Patterns in Everyforest

Patterns across a landscape are maintained through time by ecological disturbances. An ecological disturbance is any relatively discrete event that disrupts the structure of a population or community of plants and animals or disrupts the ecosystem as a whole and thereby changes the availability of resources or restructures the physical environment. Cycles of ecological disturbances, ranging from small grass fires to major hurricanes, can be characterized by their distribution in space and the size of the disruption they make as well as their frequency, duration, intensity, severity, and predictability.[6]

Be advised that changes in the patterns of North American forests before the Europeans settled were closely related to topography and to the configuration of Nature's disturbances, especially fire. When the Europeans began to alter the landscape through such introductions as the grazing of livestock and the clearing of forests for agriculture, they effectively redesigned the patterns of the forests and grasslands only selectively because these introductions accompanied human settlement and the consequent exploitation of the land. Nevertheless, those human-created disruptions began to cause unforeseen modifications in the landscape that we are now having difficulty controlling.

Although my example is taken from the forest I know best, what I am about to relate has happened in every nation in the world in some way, at some time in its history.[7] In the Pacific Northwest, for example, vast areas of connected, structurally diverse forest, of which the National Forest System was once constituted, have been fragmented and rendered homogeneous by clear-cutting small, square blocks of old trees; by converting these clear-cuts into even-aged stands of nursery stock; and by leaving small, uncut, square blocks of timber between the cut blocks. This "staggered-setting system," as it is called, requires an extensive network of roads.

So, by 1998, before half the land area was cut, 433,000 miles of logging roads—more than eight times the size of the U.S. interstate highway system—penetrated almost every water catchment.[8] And, when half the land was cut, the whole of the National Forest System became an all-of-a-piece patchwork quilt with few, if any, forested acres in any given area large enough to support those species of plants, birds, and mammals that require interior forest as their habitat.[9] Moreover, there is *nothing* ecologically friendly about a road—any road—with respect to its existence in an ecosystem.[10] A similar scenario took place in the Great Plains with fence and plow, where extensive grasslands were converted to agricultural fields and monocultural crops.

Changing a formerly diverse landscape into cookie-cutter sameness has profound implications. Nature's disturbance regimes, such as fires, floods, windstorms, and outbreaks of insects, coupled with such disruptions of human society as urbanization and pollution, are significant processes in shaping a landscape. The function of these processes is influenced by the diversity of the existing landscape pattern.

Disturbances vary in character and are often controlled by physical features and patterns of vegetation. The variability of each disturbance, along with the area's previous history and its particular soil, leads to the existing vegetational mosaic.

The greatest single perturbation to an ecosystem is the biophysical turmoil caused by humans attempting to control the size (minimize the scale) of the various cycles of Nature's disturbance with which an ecosystem has evolved and to which it has become adapted. Among the most obvious (and well-intentioned) acts to control is the suppression of fire, a practice that has allowed areas of forest to become so dense in numbers of trees that, as a growing part of the forest system, they stagnated. Coupled with this kind of human manipulation is the systematic, large-scale, exploitation of timber via clear-cut logging—a catastrophic human activity to which no forest on Earth is adapted or adaptable.

As we struggle to minimize the scale of Nature's dynamics in an ecosystem, we alter the system's ability to resist or cope with the multitude of invisible stresses to which it adapts through the existence and forces of the very cycles of disruption that we "control." Clearly, today's forest fires are more intense and more extensive than in the past because of the buildup of fuels since the onset of fire suppression. Many forested areas are now primed for catastrophic fire.[11] Outbreaks of plant-damaging insects and diseases spread more rapidly over areas of forest that have been stressed through the removal of Nature's own upheavals, to which they are adapted and which control their insects and diseases.[12]

The precise mechanisms by which ecosystems cope with stress vary, but one is tied closely to the genetic selectivity of an ecosystem's species. As an ecosystem changes and is influenced by increasing magnitudes of stresses, the replacement of a stress-sensitive species with a functionally similar but more stress-resistant species maintains the system's overall productivity. This replacement of species—backups—can result only from within the existing pool of biodiversity.[13]

Human-introduced disturbances, especially fragmentation of habitat, impose stresses for which ecosystems are ill adapted to cope.[14] Not surprisingly, biogeographical studies show that *connectivity* of habitats within and among ecosystems is of prime importance to the persistence of plants and animals in viable numbers in their respective habitats—again, a matter of biodiversity. In this sense, the landscape is a mosaic of interconnected patches of habitat that act as corridors or routes of travel between one suitable patch of habitat and another.

Whether populations of plants and animals survive in a particular landscape depends on the rate of local extinctions from a patch of habitat and the rate with which an organism can move among existing patches. Those species living in habitats isolated as a result of fragmentation are not likely to persist, especially if they are relatively sedentary, such as salamanders. Fragmentation of habitat, the most serious threat to biological diversity, is the primary cause of the present global crisis in the rate of biological extinctions. On public lands, such as national forests, much, if not most, of the fragmentation of the habitat is a direct, although unanticipated, effect of management policies (like the location of roads) that stress the short-term production of commodities at the long-term expense of the environment.[15]

Modifying the existing connectivity among patches of habitat strongly influences the abundance of species and their patterns of movement. The size, shape, and diversity of patches also influence the configuration of species abundance, and the shape of a patch may determine which species inhabit it. The interaction between the processes of a species' dispersal and the biophysical arrangement of a landscape determines the temporal dynamics of the species' populations. Populations of wide-ranging organisms may not be as strongly affected by the spatial distribution of habitat patches as are the more sedentary species.[16]

Nevertheless, adaptable landscapes with desirable productive capacities to pass to our heirs necessitate a focus on two primary things: (1) caring for and "managing" for a sustainable connectivity and biological richness among such areas as logged forest, agricultural fields, livestock pastures, and urban developments within the context of the landscape as a whole and

(2) protecting existing biodiversity—including habitats—at any price for the long-term sustainability of the ecological wholeness and biological richness of the patterns we create across landscapes.[17]

Patterns in Everycity

Pattern is also important in a city and, like that of the national forest, changes over time. In the olden days of the United States, for instance, the church was the most important building and so was centrally located in a place of prominence. With time, however, the money chase became the dominant pursuit of most citizens, which resulted in buildings dedicated to the garnering of money becoming larger and larger, springing up wherever it was most convenient and dominating the skyline—as the twin trade towers in New York City once did. As money progressively crept into the human psyche, where it represents the "real" vehicle to the illusion of material security, it began to drive the pattern of development known today as "urban sprawl." Consequently, the early sense of community was fragmented—like the world's natural habitats—and is today all but absent in most cities.

Sprawl gobbles up irreplaceable agricultural lands. Cookie-cutter houses are concentrated into developments isolated from everything else dealing with community, and the downtown areas stagnate. Speed, rather than care, creeps into the building trade as houses spring up in blocks and lines and circles, built for speculation. As speculation infiltrated the housing market, speed, sameness, and clustering became marks of efficiency and greater profit, setting the tone for the future—a tone reflected in the night sky as the once-brilliant stars disappear into a seemingly eternal mask of light pollution.

With the stage set by the housing industry in the United States after World War II, things began to change noticeably as corporate depersonalization commenced its insidious cancerlike growth into the heart of community. Roads that became bigger, straighter, and faster and increasingly went through prime agricultural land connected shopping malls. Then came larger and larger subdivisions with cheaper (but no less expensive), ticky-tacky, tract housing, some of which was constructed in floodplains and unstable soils.

Driving on superhighways became a necessity, and with it came pollution of air and water that increases with every mile that *has to be driven and every additional automobile on the road*. And, the gentle motion and the relaxed pace of the traditional street gave way to ever-increasing speed.

As author Jean Chesneaux observed: "The street as an art of life is disappearing in favour of traffic arteries. People drive through them on the way to somewhere else." Unfortunately, there is no word in the English language with a positive connotation for going slowly or lingering on streets as a way to participate in a sense of community.

People started losing their sense of connection as centralization within urban sprawl increasingly specialized the human landscape, and communities began falling apart. A sense of place—of a familiar, friendly community, where everyone left their homes and cars unlocked—gave way to a sense of location, as more and more people became transients. Many were, and still are, moved about at corporate or agency will.

Within this context, zoning that is not tied to a community vision or plan has been largely detrimental in the sense that it is often used politically to foster such things as concentrated shopping malls and isolated housing developments, often in irreplaceable agricultural or forested land. (See Gary Gardner, "Preserving Agricultural Resources," for a synopsis of what is happening to agricultural land.[18]) Today, in fact, more people are in jail in the United States than work full time on the nation's farms.[19] Such centralized specialization disrupts the ability of people to meet their daily requirements within a convenient area. Although convenience may not, on the surface, seem of much importance, it is in at least two ways.

In many small communities in the United States, Germany, France, Malaysia, Japan, and elsewhere, in which people can fulfill their daily requirements within walking distance of their homes, they get exercise and simultaneously keep the air cleaner by leaving their cars parked. This convenience, this *centralized generalization* of goods and services, also allows people to meet one another on the street and in the shops, where they become familiar and often stop to exchange pleasantries. There are also places, like a particular cafe or soda fountain, a certain shade tree, or a community well, where local people daily gather and visit.

As people get to know one another, they become acquainted with what others do and on whom they can rely when in need of help. Because people know one another, they look out for one another in a free exchange of mutual caring. Of course, there are social problems and interpersonal conflicts, but mutual well-being is a great incentive to work them out.

With the advent of fragmentation through *centralized specialization*, people are increasingly forced to drive from their homes to go to the grocery store, the doctor, the post office, the hair stylist, and so on. This forced reliance on an automobile not only isolates people from the opportunity of getting to know one another by walking familiar streets together but also isolates them in the shopping malls, where people are in essence visitors from foreign housing developments. And, because such centralized specialization necessitates driving automobiles, it dramatically increases the problem of polluting our common air, soil, and water.

Gone are most of the friendly communities of old, those centered around daily living within walking distance, where open spaces from the surrounding landscape were integrated within the community itself. Increasingly, today's communities are what might be called "suburban block communities," where a few neighbors may be friendly with one another, at times almost as a defense against the larger world.

People tend to defend these small enclaves, often like special interest groups defending an intellectual position against an opposing point of view.

Other people cluster within fortified "communities," devoid of children and pets, surrounded by high walls and locked gates (at times with guards) to *protect* themselves from "those" on the outside. This perceived need to protect oneself is a sign of increasing fear that fosters neighborhood isolationism in the form of social disintegration instead of the overall connectivity prerequisite for real community.[20] Along with the social-environmental fragmentation of landscapes and cityscapes comes the increasing vulnerability of the biophysical infrastructure that governs them both.

Summation

Chapter 6 discussed the patterns we humans are creating across our shared landscapes. These patterns give visible expression to the heretofore-unseen desires that reside in our human psyche. Nature's configuration is consummated through the biophysical attributes and processes that ultimately constitute life, whereas humanity's design culminates in the material manifestation of an inner desire conveyed to the outer world through the medium of language and increasingly through economics. It is therefore critical to keep in mind that the spatial and temporal integrity of landscape patterns is important to account for because it is the relationship of pattern (not of numbers) that confers stability on systems, in dealing with both Everyforest and Everycity. This said, systems' stability is mediated through the interrelationships of infrastructure and is the subject of Chapter 7.

Notes

1. Ralph Waldo Emerson. *Essays*. Houghton Mifflin and Co., Boston (1876).
2. Wendell Berry. *Sex, Economy, Freedom, and Community*. Pantheon Books, New York (1993).
3. Ralph Waldo Emerson. http://www.quotationsforinspiration.com/help.htm (accessed January 20, 2009).
4. William James. http://www.educationworld.com/a_lesson/TM/WS_back_to_school_quotes.shtml (accessed January 20, 2009).
5. Margaret Mead. http://www.quotationspage.com/quote/33522.html (accessed January 20, 2009).
6. Monica G. Turner. Landscape ecology: The effect of pattern on process. *Annual Review of Ecological Systems*, 20 (1989):171–197.
7. Chris Maser. *Sustainable Forestry: Philosophy, Science, and Economics*. St. Lucie Press, Delray Beach, FL (1994).

8. Associated Press. Logging-road problem bigger than thought. *Corvallis Gazette-Times* (January 22, 1998).
9. Jerry F. Franklin and Richard T.T. Forman. Creating landscape patterns by forest cutting: Ecological consequences and principles. *Landscape Ecology*, 1 (1987):5–18.
10. W. Schmidt. Plant dispersal by motor cars. *Vegtatio* 80 (1989):147–152; David K. Person, Matthew Kirchhoff, Victor Van Ballenberghe, and others. *The Alexander Archipelago Wolf: A Conservation Assessment*. Pacific Northwest Research Station, Portland, OR (1996). USDA Forest Service General Technical Report. PNW-GTR-384; W. J. Elliot, C. H. Luce, R. B. Foltz, and T. E. Koler. Hydrologic and sedimentation effects of open and closed roads. Blue Mountain Natural Resources Institute, La Grande, OR. *Natural Resource News* 6 (1996):7–8; Richard T. T. Forman and L. Alexander. Roads and their major ecological effects. *Annual Review of Ecology and Systematics*, 29 (1998):207–231; Stephen C. Trombulak and Christopher A. Frissell. Review of Ecological Effects of Roads on Terrestrial and Aquatic Communities. *Conservation Biology*, 14 (2000):18–30; Richard T. T. Forman. Estimate of the area affected ecologically by the road system in the United States. *Conservation Biology*, 14 (2000):31–35; Jonathan L. Gelbrad and Jayne Belnap. Roads as conduits for exotic plant invasions in a semiarid landscape. *Conservation Biology*, 17 (2003):420–432.
11. Michael Williams. *Americans and Their Forests: A Historical Geography*. Cambridge University Press, New York (1989); Stephen J. Pyne. *Fire in America: A Cultural History of Wildland and Rural Fire*. University of Washington Press, Seattle, WA (1997); Stephen J. Pyne. *Year of the Fires: The Story of the Great Fires of 1910*. Viking Press, New York (2001).
12. Timothy D. Schowalter. Adaptations of insects to disturbance. *In*: *The Ecology of Natural Disturbance and Patch Dynamics*, S. T. A. Pickett and P. S. White (eds.), 235–386. Academic Press, New York, 1985; Timothy D. Schowalter, W. W. Hargrove, and D. A. Crossley Jr. Herbivory in forested ecosystems. *Annual Review of Entomology*, 31 (1986):177–196; Timothy D. Schowalter and Joseph E. Means. Pest response to simplification of forest landscapes. *Northwest Environmental Journal*, 4 (1988):342–343; Timothy D. Schowalter and Joseph E. Means. Pests link site productivity to the landscape. *In*: *Maintaining the Long-Term Productivity of Pacific Northwest Forest Ecosystems*, David A. Perry, R. Meurisse, B. Thomas, R. Miller, and others (eds.), 248–250. Timber Press, Portland, OR, 1989; David A. Perry. Landscape pattern and forest pests. *Northwest Environmental Journal*, 4 (1988):213–228; Timothy D. Schowalter. *Insect Ecology: An Ecosystem Approach*. Academic Press, San Diego, CA (2000).
13. David J. Rapport, H. A. Regier, and T. C. Hutchinson. Ecosystem behavior under stress. *American Naturalist*, 125 (1985):617–640; David J. Rapport. What constitutes ecosystem health? *Perspectives in Biology and Medicine*, 33 (1989):120–132.
14. Rapport et al. Ecosystem behavior; Rapport. What constitutes ecosystem health?
15. Chris Maser. *Earth in Our Care: Ecology, Economy, and Sustainability*. Rutgers University Press, New Brunswick, NJ (2009).
16. Ibid.

17. The foregoing discussion of a forest is taken from Chris Maser. *Ecological Diversity in Sustainable Development: The Vital and Forgotten Dimension*. Lewis Publishers, Boca Raton, FL (1999).
18. Gary Gardner. Preserving agricultural resources. *In*: *State of the World 1996: A Worldwatch Institute Report on Progress Toward a Sustainable Society*, Lester R. Brown, Janet Abramovitz, Chris Bright, and others (eds.), 78–94. W. W. Norton and Co., New York, 1996.
19. Colin Tudge. Enlightened agriculture. *Resurgence*, 221 (2003):18–21.
20. The foregoing discussion of my hometown is taken from Chris Maser. *Sustainable Community Development: Principles and Concepts*. St. Lucie Press, Delray Beach, FL (1997).

7

Infrastructure

> Only a people serving an apprenticeship to nature can be trusted with machines.
>
> **British philosopher Herbert Read**[1]

Narrowly speaking, infrastructure refers to the basic systems and facilities serving a city, county, or nation. Generally speaking, infrastructure is that part of Everyforest and Everycity that simultaneously allows the components to interact and facilitates their interaction in a functional, systemic manner. Infrastructure, in both the restrictive and the broad sense, functions as a means of transferring energy from one part of a system to another. In essence, infrastructure is a system of energy interchange composed of subsystems of energy interchange, all of which come from the Earth's natural wealth—not the monetary wealth invented by humans.

Infrastructure is the interplay among (1) Nature's biophysical principles and (2) society's attempt to impose its will on Nature's biophysical principles in order to create a design of choice, often at the cost of social-environmental principles of sustainability.

The Energy Interchange System

To understand the role of infrastructure as an energy interchange system, imagine that you are videotaping a pebble dropping into a pool of quiet water. As the pebble hits water, concentric rings radiate outward from the point of contact, becoming wider and more diffuse the farther from the epicenter they travel. If the pool is large enough and your lens is strong enough, you could capture the visual disappearance of the rings but not their ultimate dissipation. Now, if you play the videotape backward, you would see the rings coming together, getting crisper and spatially closer as they approach the place where the pebble struck the water's surface.

The foregoing is an exceedingly simplistic look at the intersecting rings. To more clearly visualize what is really going on, you would have to videotape a hundred pebbles being dropped into a very large pool at different places and different times within a two-minute period as well as from various heights above the water. Watch their rings intersect again and again and again. Playing the tape forward and backward would give you a better mental image of what is happening within and between Everyforest and Everycity. This image, however, is confined to horizontal rings, while reality is a dynamic of continually intersecting rings going simultaneously in all directions through space and time.

Here, it is good to remember that the *first law of thermodynamics* states that energy can neither be created nor destroyed, meaning that, while energy can be changed in form, distribution, and function, the quantity remains the same. A graphic, although perhaps not an entirely accurate, demonstration of this law can be made by playing the videotape forward and then backward. Note the dissipation of energy as the ripples weaken in contour the farther they extend from the epicenter of the pebble's collision with the water. Then, note the reverse phenomenon as the ripples strengthen in like measure the closer to they get to the epicenter of the pebble's impact. How might this concept work in Everyforest and Everycity?

Energy Interchange in Everyforest

While reading about the energy interchange in a forest, keep the trees in mind as the representative center of the concentric rings in the pebble-pool interface. Although a city draws its energy from near, far, and wide, a forest tends to have much of its energy drawn from relatively near, with the obvious exception of that derived from the sun's light. Here, I must point out that following the ins and outs of the energy interchange within a forest is complicated at best. It took me over thirty years of research to comprehend what is in this section, and there is still much that I neither know nor fully understand. So, please be patient with yourself and with me as you wend your way through the energy labyrinth of the forest.

All things in Nature are neutral when it comes to any kind of human valuation. Nature has only intrinsic value in that each component of a forest, whether a microscopic bacterium or a towering 800-year-old tree, develops its own structure, carries out its own function, and interacts with other components of the forest through interdependent processes and feedback loops. No component is more or less valuable than another; each may differ in form, but all are complementary in function.

To introduce you to the notion of energy interchange in a forest, picture a single, large tree in the middle of a slicked-off clear-cut. This tree is vulnerable to the energy of the wind in that it may get blown over in a storm or struck by lightening but is less vulnerable to death through a forest fire because there is little around it to burn. In contrast, a tree in the middle of a forest is supported against the power of a storm wind by the collective strength of the surrounding trees. But at the same time, it is more vulnerable to death by fire because there is a sea of treetops that, once aflame, can burn with incredible spread and intensity. This brief scenario illustrates that nothing in the universe is entirely free because everything is defined by its cause-and-effect relationship to everything else—seen and unseen.

Now, let us tackle a more complex depiction of the energy interchange in a forest. Most higher plants worldwide—from the northern tundra, to the desert, to the tropical rainforest—have an obligatory, symbiotic relationship with certain fungi that are central to their processes of capturing nutrients from the soil. In fact, fossil evidence of the earliest terrestrial plants, as well as current molecular studies, shows that plant roots and certain fungi coevolved as symbiotic partners to form structures known as *mycorrhizae*, a word from the Greek *mykes* (a fungus) and *rhiza* (a root). Mycorrhizae are generally distributed throughout present-day plant communities.

The hyphae, or individual fungal strands, each one to two cells thick, form the main structural elements of the mycorrhizal fungi. ("Hypha," singular, comes from the Greek *hyphe*, a web.) The hyphae either penetrate the cells of a plant's roots to form an *endomycorrhiza* (*endo*, meaning mycorrhiza *inside* the root) or ensheath the root to form an *ectomycorrhizae* (*ecto*, meaning mycorrhiza *outside* the root).

In the nutrient-deficient conditions of most forests, ectomycorrhizal fungi colonize at least 90 percent of a tree's "feeding" roots. This colonization forms a layer of fungal tissue or "mantle" around the tree's feeding root, thereby creating an interface between the tree's feeding root and the soil. From this mantle, individual hypha aggregate and organize into rootlike structures that grow out from the feeding root into the soil, where they act as an extension of the tree's root system. The aggregate of hyphae is also referred to as *mycelia* (singular, mycelium), meaning "made of mushrooms" in New Latin. A mycelium is a mass of hyphae that forms the nonreproductive part of a fungus.

Some mycorrhizal fungi are host specific, meaning that a given fungus colonizes a single species of plant, whereas others are generalists and colonize a number of host plants. Douglas-fir or birch, for example, can be colonized by many species. In turn, these fungi extend from tree to tree and so form linkages among trees. The generalized nature of host compatibility ensures that almost all trees and many other plants in an undisturbed forest ecosystem, regardless of species, are interconnected by billions of miles of hyphae, organized into systems of mycelia, that originate from a diverse population of mycorrhizal fungi. Consequently, if you could pull up a whole forest and gently wash the soil from the roots of the plants, you would find a net of mycelia connecting the entire forest—one of the reasons an old forest is so retentive of its soil nutrient capital.[2]

Through the obligatory, symbiotic mycorrhizal association, a plant, such as firs or eucalyptus, provides carbohydrates to its mycorrhizal symbiont, while the fungal symbiont mediates the plant's uptake of nitrogen, phosphorus, other minerals, and water. In addition, the mycorrhizal association promotes the development of fine roots; produces antibiotics, hormones, and vitamins useful to the host plant; protects the plant's roots from pathogens and environmental extremes; moderates the effects of heavy metal toxins;

and promotes and maintains soil structure and the forest food web. On the other hand, this association is expensive, with an estimated 50 to 70 percent of the host plant's net annual productivity being translocated to its roots and associated mycorrhizal fungi.[3]

When access to nutrients is increasingly restricted in a forest ecosystem that is already nutrient impoverished, mycorrhizal fungi can influence both the interactions among plants and the species composition of the plant community itself. It appears there is an additive beneficial effect that comes with each species of mycorrhizal fungus colonizing a given plant, which in turn could mean an increase in both biodiversity and ecosystem productivity with a greater number of fungal symbionts. This scenario seems likely because experimentation has shown that as the number of fungal symbionts increases so does the collective biomass of roots and shoots as well as the species diversity of the plant community. Conversely, as a forest is disturbed through exploitive forestry and the use of artificial fertilizers, the function of the mycorrhizal system can be impaired.[4]

In addition to gleaning nutrients from the soil and translocating them into the host plant, mycorrhizal fungi, along with roots of the host plant and the free-living microbial decomposers in the soil, are significant components of the global balance of carbon. Much of the carbon balance is mediated by photosynthesis and drives the respiration or "breathing" of the soil. Photosynthesis in turn is the synthesis of complex organic materials (especially carbohydrates from carbon dioxide, water, and inorganic salts) by using sunlight as the source of energy, with the aid of chlorophyll and its associated pigments. The "photosynthates" (equals "nutriments") produced by the process of photosynthesis (sent from the green, aboveground portion of a plant to its roots and mycorrhizal symbionts) are critical in maintaining the soil's respiration, through which carbon is extracted from the soil. Although the photosynthetic carbon helps balance the loss of soil carbon through respiration, the production of photosynthates is mediated more by annual seasonality than by the temperature of the soil.[5]

Keeping the above in mind, let us investigate a self-reinforcing feedback loop in the coniferous forests of the Pacific Northwest in which Douglas-fir and western hemlock predominate in the canopy of the old forest. Herein lives the northern spotted owl, which preys on the northern flying squirrel as a staple of its diet. The flying squirrel in turn depends on truffles—belowground fruiting bodies of ectomycorrhizal fungi.[6]

Flying squirrels dine heavily on truffles and, having eaten a truffle, defecate live fungal spores onto the forest floor, which, on being washed into the soil by rain, inoculate the roots of the forest trees.[7] As stated earlier, these fungi depend for survival on the live forest trees to feed them sugars produced in the green crowns of the trees. In turn, the fungi form extensions of a tree's root system by collecting minerals, other nutrients, and water that are vital to the tree's survival. The fungi also depend on large, rotting trees lying on and buried in the forest floor for water and the formation of humus in the soil. *Humus*, which lends soil its dark color, is the Latin word for "the ground, soil" or alternatively the New Latin word *humos*, meaning "full of earth." Further, nitrogen-fixing bacteria occur on and in the ectomycorrhiza, where they convert atmospheric nitrogen into ammonia, a form that is usable by both fungus and tree.[8] All of these organisms are conduits of energy derived in one part of the forest, transformed within their respective bodies, and introduced in a different form into another part of the forest.

Such small mammal-fungal-tree relationships have been documented throughout forests of the United States (including Alaska) and Canada.[9] They are also known from forests in Argentina,[10] Europe, and Australia.[11]

To add to the overall complexity of a late-successional, indigenous forest, a live old tree eventually becomes injured or sickened with disease and begins to die. How a tree dies determines how it decomposes and reinvests its biological capital (organic material and chemical elements) back into the soil and eventually into the next forest.

A tree may die standing as a snag to crumble and fall piecemeal to the forest floor over decades, or it may fall directly to the forest floor as a whole tree. How a tree dies is important to the health of the forest because its manner of death determines the structural dynamics of its body as habitat. Structural dynamics in turn determine the biochemical diversity hidden within the tree's decomposing body as ecological processes incorporate the old tree into the soil from whence it came and the next forest must grow.

What goes on inside the decomposing body of a dying or dead tree is one example of the hidden biological and functional diversity that is totally ignored in a typical economic valuation of a forest. Consequently, that trees continue to become injured, diseased, and die is critical to the long-term biophysical health of a forest because the last is an interactive, organic whole defined not by the pieces of its body but rather by the interdependent, functional relationships of those pieces creating the whole—the intrinsic value of each piece and its complementary function.

Regardless of how it dies, the snag or fallen tree is only an altered state of the live tree; so, the live large tree has to exist before there can be a large snag or large fallen tree. A basic problem inherent in understanding this scenario lies in the fact that few of the world's leading scientists, legislators, policy makers, or business executives could name the five classical kingdoms of Nature (bacteria, fungi, algae, plants, and animals) or how they complement one another. This dearth of ecological literacy shows a basic lack of understanding how ecosystems and their self-reinforcing, biophysical feedback loops work, as well as a basic lack of understanding of the long-term consequences of short-term political and economic decisions.

Finally, the various functions of the individual species, when melded in the collective, form a self-reinforcing feedback loop of mutually dependent interrelationships in which the spotted owl preys on the flying squirrel; the flying squirrel eats the truffle; the fungus is closely associated with large wood on and in the forest floor for water and the formation of humus (hence, fungus, spotted owl, and flying squirrel are all dependent, either directly or indirectly, on the same large decomposing wood on the forest floor); the fungus feeds the tree; the tree feeds the fungus; the squirrel inoculates the tree's roots with fungal spores that help keep the tree healthy; the tree houses the squirrel and the spotted owl; the owl eats the squirrel, and on and on ad infinitum.[12] Throughout this entire suite of processes, sunlight (solar radiation) is the only new source of energy (true investment of new energy) that drives the living portion of the forest's biophysical systems, while wind and water are the main sources of massively destructive energy that helps to maintain the infinite novelty of which living systems are a manifestation—Everyforest.

From a human perspective, there are two indispensable items that flow out of a forest on the concentric rings of pebble and pond. One is oxygen given off by the trees and other chlorophyll-containing vegetation in exchange for the intake of carbon dioxide. The other is water, as previously discussed, albeit, the water available for use in cities is the excess of that required by the forest itself.

In discussing oxygen, note that it originates within the green plants of the forest and travels outward on the concentric rings. In contraposition, water comes into the forest as rain or snow and is used by the forest and transpired back into the air; the excess soaks into the soil only to gather in rill, stream, and river, where it is siphoned off for use in the city.

Energy Interchange in Everycity

Energy flows into a city from the outer reaches of the world. That energy comes in many forms:

- Oil, derived from living organisms over epochs of geological time, is today used directly to heat buildings and lubricate machinery as well as converted to kerosene, diesel, and gasoline that fuel automobiles, tractors, tanks, trains, aircraft, and ships, to name a few items.
- Natural gas, produced by the decomposition of organic material from dead plants and animals and stored below ground through the millennia, is used for heating buildings, cooking food, and in some cases, as a substitute for gasoline.
- Coal, a derivative of plants wherein the sun's energy has been entombed for eons in the dark recesses of the Earth, is used to heat homes and is transformed into electricity.
- Food grown on farms, where it harvests the sun's energy, feeds us all.
- Water, evaporated from the oceans of the world, is dropped as rain and snow on the land and is piped into the city, where it quenches the thirst of plant and animal alike—including us.

- Wood, grown in forests, is used for heating, cooking, and making paper on which ideas—the energy of the mind—are transferred from one person, one generation to another.
- Manure from livestock is used in many of the poor regions of the nonindustrialized world for heating and cooking.
- Wind, the breath of the world, blows at will through cities as the gentle breezes of spring and summer and the howling voice of winter's storms; as well, it brings the rain and snow to the land from whence all life terrestrial derives freshwater.
- Nuclear fission involves splitting the nucleus of an atom in order to release its stored energy.
- Electricity—oil, gas, coal, nuclear fission, wood, water, wind, and sunlight, transformed into electrical energy—is brought into cities and used for myriad forms of work, from driving engines to lighting and heating homes.
- Human labor, the energy of mind and body, although once of priceless value, is now increasingly being replaced by machines that neither think nor feel, that have neither a sense of moral values nor the moral consciousness that accompanies them.
- Sunlight is the unifying factor, the ultimate source of energy that is simultaneously the catalyst for all the other forms of energy and stored in all the other forms of energy.

Each type of energy constitutes a specific currency in the bank of Nature's wealth. I say a "specific currency" because each allows a certain spectrum of work to be done, but none is totally interchangeable in what it allows because each functions differently (e.g., water cannot burn and wood cannot flow, just as a human can digest animal and vegetable food as bodily fuel but not coal). Accordingly, each form of currency complements the others and allows humans diverse ways to perform work.

Energy, coming from afar, is concentrated in cities in all its various forms: some from deep in the Earth (oil, natural gas, geothermal, and coal), some flowing over its surface (water), some flowing above its surface (wind), some growing out of its surface (food, wood, manure from livestock), and some radiating from the heavens (sunlight). To see how this concentration functions, run the videotape depicting the concentric rings of pebble and pond in reverse and imagine each source of energy riding the crest of a ring of gathering power as they all coalesce in the center—the city.

Whereas energy is said to be "produced" before the various forms of energy are brought into the city, that same energy is said to be "consumed" once it is in the city. In fact, there is an electrical utility in western Oregon named Consumer's Power, which is totally misleading because energy is neither produced nor consumed; it is only converted from one form to another.

When, for instance, the ancient energy of the sun's light, stored for eons in today's fossil fuels, is released (consumed), creating thermal, mechanical, or electrical energy, only the form of the energy has changed, bringing up the *second law of thermodynamics*. This law states that the amount of energy in forms available to do useful work can only diminish over time, thus representing a diminishing capacity to do useful work in a city—in every city.

When dealing with the notion of social-environmental sustainability, it is vital to understand that an "expenditure" of energy means to convert a useful *and* available form of energy (that with which work can be done) to a less-useful or available form. It is also vital to understand that some forms of energy have problems of pollution associated with them when converted from one form to another—the conversion of fossil fuels to electricity or gasoline is illustrative.

In the process of converting coal to electricity, pollutants (nitrogen oxides, sulfur dioxide, carbon monoxide, etc.) are spewed into the air and carried hundreds or thousands of miles from the coal-fired power plants that produced them in a phenomenon known as acid deposition, commonly called *acid rain*. This pollution of air (which is part of the global commons) represents the concentric circles of pebble and pond radiating outward from the center, from a city into the wider world—the videotape played forward.

In turn, air pollution directly affects vegetation by altering the quality of the soil and water as well as the quality and quantity of the sunlight that drives the plant/soil processes. The chemicals we dump into the air also alter the climate and thus the environment in which the vegetation grows.

Acid rain has long been recognized as a pollution problem in Europe, where statues and gargoyles that once proudly adorned city streets and plazas and guarded centenarian buildings have had their faces dissolved over recent decades. This phenomenon is not confined to cities in Europe or Asia, however. It is also found in forest and fen, in highland and lowland. There it also is destroying the essence of life as it joins league with other forms of industrial and technological pollution, where it contributes to a phenomenon the Germans call *Waldsterben*, the dying forest. *Waldsterben* raises the question: What happens when outgoing and incoming rings intersect?

Intersecting Rings

As the rings of cause flow outward from the collision of pebble and water, they become the rings of effect. In turn, each ring of effect becomes a cause when it intersects with a ring of effect from a pebble dropped in another part of the pool. In this way, each cause becomes an effect that becomes a different, often a synergistic, cause that initiates still another effect and so on ad infinitum. And, we humans are constantly adding to the synergistic interactions of the intersecting rings with each substance we introduce into the environment—a substance immediately out of our control. But in this case, I discuss intersecting rings and cities first because they have a profound effect on forests.

Intersecting Rings and Cities

The intersecting rings with which people are most familiar pertain to changes in the global climate, which are causing the planet's temperature to fluctuate. The nonhuman causes include sulfate droplets, called *aerosols*, spewed aloft by erupting volcanoes, which cool the atmosphere by reflecting the sun's light as they ride air currents into cities. On the other hand, the human influence stems mostly from emissions of industrial gases, like rings of carbon dioxide and sulfate aerosols flowing out from the industrial smokestacks of cities, thereby redistributing heat in the atmosphere.[13]

Precisely how much of this global warming is due to humans is unknown, but there is little doubt that we are affecting the process in some surprising ways. For instance, tiny particles of industrial pollution ride the currents of air as they leave major industries, where they intersect the inflowing rings of weather and stifle precipitation in many regions of the world by affecting clouds that produce rain and snow.[14] These minute particles cause physical changes in clouds that prevent water from condensing into raindrops or snowflakes.

Air pollution can eliminate precipitation from certain types of clouds, indicating that human activity is potentially altering clouds and thus worldwide patterns of precipitation. Although numerous factors influence precipitation, changes in the weather can spread hundreds of miles downwind of a large industry's plume of polluted air, such as electric utilities.[15] According to the U.S. Environmental Protection Agency, electric utilities annually dump more than 1.1 billion pounds of toxic chemicals into the environment—about 70 percent of it into the air through power plant smokestacks, mostly from coal-burning plants.[16]

It is a physical process, rather than a chemical one, that blocks the formation of rain and snow. Industrial plants spew pollutants into the air (outflowing rings) that are formed by the combustion of fuel; the particles are much smaller than the water droplets normally found in clouds (inflowing rings). These tiny particles inhibit the cloud's droplets of water from coalescing into drops large enough to fall as rain. Small droplets of water are also slow to freeze, which reduces the ice crystals in clouds. In short-lived clouds, the lack of large droplets can reduce, or even eliminate, precipitation.[17]

Intersecting Rings and Forests

As this juncture, the city-forest relationship gets more complicated in that the outflowing rings of carbon dioxide and chemical pollutants from the city become the inflowing rings of the forest, even as the outgoing rings of the forest (oxygen and water) form the incoming rings of the city. In terms of the forest, this intersection is both positive and negative; while the forest trees use carbon dioxide in the photosynthetic process of their crowns, the pollutants simultaneously enter the soil, where they instigate the declining health of the same trees through their roots.

The dying-forest syndrome, mentioned earlier, is not exclusively the property of Europe. It is found in every industrial country, including the United States and Canada. Here, called *forest dieback*, it manifests itself primarily along the eastern seaboard, where declining growth rates and the demise of red spruce and other species of trees, particularly at high elevations, are attributed to atmospheric pollution, of which acid deposition is one of the most widely spread components.

A primary human source of the precursors to acid deposition are the coal-fired power plants that still provide more than 55 percent of the electricity generated in the United States and account for about one-third of the nitrogen oxides and about two-thirds of the sulfur dioxide produced each year.[18] These atmospheric pollutants are capable of a phenomenon known as *long-distance transport*, which simply means that they can travel great distances from their sources on air currents (even as far away as the Arctic and Antarctica) before being deposited on agricultural fields and forests, where they affect plant growth.

Direct effects of acid deposition include reduced functioning of plant roots that may be caused by chemical changes in the soil as a result of acid deposition. Here, the acid reduces translocation of carbohydrates either from pollution-damaged shoots or from excessive nitrogen. In fact, the emission of pollutants has tripled the level of nitrogen some forests are receiving.

Change in the level of nitrogen works as follows: People are accustomed to the idea that if one puts nitrogen on a cornfield, the corn grows faster, but then corn happens to be a plant that opens its stomata (pore-like entrances into the leaves) and takes up more carbon dioxide to create a balance. Thus, up to some point, the more nitrogen corn is given, the more carbon it will acquire, and the better it will grow. Yet, unlike corn, most forest species will not grow any faster with additional nitrogen.

Since trees do not take up any more carbon to offset the additional nitrogen, an altered carbon-nitrogen ratio develops in the plant tissue that means trees are receiving three times the level of their nitrogen tolerance. This *does not* mean that they are getting three times their optimum level but rather three times their *tolerance*. Such a shifted carbon-nitrogen ratio translates into, among other things, an alteration of the materials plants produce to resist diseases and insects (a weakened immune system), permitting pathogenic fungi (those that create rot in living trees) to enter where heretofore they had been excluded.[19]

The problem of acid deposition in forests is worst in spring when snow melts and releases into streams the spike of acidity it has accumulated and stored over the winter. Although streams have their highest acidic levels of the year in spring, some are more affected by the spike in acidity than others. This acidity is now an outflowing ring of effect from the forest, which becomes an inflowing ring of cause that reduces water quality in the city. While the affect of such a flush of acid is perhaps most apparent in streams, it also affects soil. In fact, how much a particular stream is affected by acidification depends on two primary factors in the soil of the water catchment.

The first factor affecting acidification in streams is how long the water is retained in the soil, where it comes in contact with soil particles that neutralize its acidity. Like steeping tea, the longer the water is in contact with the soil, the more its acidity is neutralized. In other words, given enough time, soil rich in calcium and other so-called base cations can exchange these ions for positively charged hydrogen ions, which make rainwater and melting snow acidic. Soil with a good capacity to buffer acidity has an abundance of these base cations, whereas soil with a poor capacity to buffer acidity does not.

Soil's natural capacity to buffer acidity is derived from, and maintained by, the slow weathering of calcium and other base cations from minerals in the inorganic parent materials. But, when soil with a poor buffering capacity is unable to counteract acidity, aluminum and other metals leach into streams. When these metals are in a high concentration, they become poisonous to aquatic life. In the northeastern United States, for instance, acid rain and melting snow are depleting the calcium and other acid-neutralizing substances faster than they can be replenished by the weathering of parent materials in the soil.[20]

One effect of acidity in the soil is the reduced tolerance of red spruce to cold temperatures. Although red spruce is generally hardy enough to survive in the coldest forests of the northeastern United States, such environmental disruptions as the depletion of calcium in the soil due to acid deposition reduce the ability of the tree to tolerate cold during severe winters. At such times, a significant percentage of its foliage is killed. Persistent injuries due to cold temperatures are a leading contributor to the decline of red spruce along the eastern seaboard.

Damage to red spruce appears to be greatest at high elevations and in those western portions of the region that are closest to the source of the sulfur and nitrogen pollution. When extensive mortality of the newly formed buds accompanies the damage of the previous year, the combined loss of the most productive foliage, coupled with a reduced capacity for new growth, is likely to dramatically alter the biological characteristics of surviving red spruce forests.[21]

Again, to visualize more clearly energy exchange through the idea of intersecting rings, put a new tape in your video camera and film a hundred pebbles being dropped into a very large pool at difference places and different times within a two-minute period as well as from various heights above the water. Watch their rings intersect again and again and again. Playing the tape forward and backward will give you a better mental image of what is happening between forests and cities. Again, this image is confined to horizontal rings, while reality is a dynamic of continually intersecting rings going simultaneously in all directions through space and time. Some effects of the intersecting rings in both forests and cities are at least partially guided by conduits in which the processes of transmission are hidden from view.

Conduits of Interchange

Although technically a *conduit* is a hollow tube of some sort, as I said earlier, I use the term to connote any system employed specifically for the transfer of something from one place to another. Every living thing, from a virus to a bacterium, fungus, plant, insect, fish, amphibian, reptile, bird, or mammal, is a conduit for the collection, transformation, absorption, storage, transfer, and expulsion of energy. In fact, the function of the entire biophysical system is tied up in the collection, transformation, absorption, storage, transfer, and expulsion of energy—one giant energy-balancing act.

Although living things have all evolved with conduits in the form of tubules, tubes, veins, intestinal tracts, and so on that clearly fit the technical notion of a conduit as some kind of hollow tube, not everything in the biophysical system that transfers energy is hollow. Both a telephone wire and an electrical wire are solid, yet both transfer energy and so can be thought of as a conduit, one for sound waves and the other for electrical pulses. And then, there are streams and rivers that conduct rushing water from one place to another and rocks that receive energy from the sun, store it during the heat of day, and release it again in the cool of night. With this in mind, let us visit a forest.

Conduits in Everyforest

Conduits in a forest range in size from a one- to two-cell-wide fungal hypha that transmits nutrients from the soil into the roots of a forest tree to the channel of a major river, which can be more than a mile (one and a half kilometer) wide. There are also termite tunnels in wood, mammal tunnels in soil, cellular conduits for the passage of water and nutrients within the tissue of plants, belowground rivers, mouse runways through the herbaceous groundcover, wildlife migration corridors, and riparian zones, to name a few.

In addition, the decomposing woody roots attached to the stumps of old, "grandparent" trees compose a "plumbing system" of sorts. This system of decomposing tree stumps and their large, deep, hollow roots frequently forms interconnected, surface-to-bedrock channels that rapidly drain water from heavy rains and melting snow. As roots slowly rot, they collapse and plug the channels, forcing more water to drain through the soil matrix, thereby reducing its cohesion while increasing the hydraulic pressure and the probability of mass soil movements (or landslides).

These plumbing systems cannot be replaced by the young trees of modern forests because their roots are too small and shallow to form the conduits necessary to rapidly move large quantities of water deep into the soil. Local extinctions of the biophysical functions performed by grandparent trees, alive and dead, seems to pass unnoticed into the shadow realm of bygone forests.

Then, of course, there are all of the animals themselves, whose multiple functions and their effects are neither as obvious nor as measurable in the forest as they are in the city. To illustrate, think about the fecal pellets of the California red-backed vole, northern flying squirrel, and the deer mouse. Why on earth, you might wonder, would I use fecal pellets as a vehicle of explanation? They are, after all, the final product to exit these three mammalian conduits.

A fecal pellet is more than a package of waste products; it is a "pill of symbiosis" dispensed throughout the forest. Each fecal pellet contains four components of potential importance to the forest: (1) spores of mycorrhizal fungi, (2) yeast, (3) nitrogen-fixing bacteria, and (4) the complete nutrient component for nitrogen-fixing bacteria.

Each fecal pellet contains viable spores of the previously discussed mycorrhizal fungi. Each also encapsulates the entire nutrient requirement for nitrogen-fixing bacteria, as well as "antifreeze" that protects the bacteria during the cold of winter. Without the antifreeze, the bacterial cells would freeze, then rupture and die when feces deposited during winter thaw in spring.

Viable nitrogen-fixing bacteria, yeast, and spores of mycorrhizal fungi all survive passage through the digestive tracts of these rodents. In addition, fecal pellets contain the complete compliment of nutrients necessary for the nitrogen-fixing bacteria. These relationships have several implications for forest habitats.

The yeast, as a part of the nutrient base, has the ability to enhance growth and nitrogen fixation in the bacteria. Abundant yeast propagules may also augment fungal germination because spores of some mycorrhizal-forming fungi are stimulated by extractives from other fungi, such as yeast.

Inoculation of soil with organisms carried in the feces of rodents is probably common in forest ecosystems worldwide. The burrowing red-backed voles and tree-dwelling flying squirrels are obligate forest dwellers. The nitrogen-fixing bacterium *Azospirillum* spp. (no common name) contained in the feces of these two rodents not only can penetrate the roots of plants but also appears to have considerable longevity in soil under some conditions.

The deer mouse, on the other hand, both lives within the forest and is one of the first small mammals to occupy clearings after logging or fire (even soil that has been severely altered by a hot fire), so it can reinoculate the soil with viable nitrogen-fixing bacteria, yeast, and spores of mycorrhizal fungi. Although the fungal spores may not survive high surface temperatures in openings, they could survive under large fallen trees on the surface of the soil, where deer mice are active, or below the surface of the soil in rodent burrows. Unlike the fungal spores, the nitrogen-fixing bacterium *Clostridium butyricum* (no common name), found in the feces of deer mice, has a built-in survival mechanism (an endospore wherein the bacterium is dormant) by which it can withstand temperatures up to 176° Fahrenheit (80° Celsius).

The fate of fecal pellets varies, depending on where they fall. In the forest canopy, the pellets might remain and disintegrate in the treetops, or a pellet could drop to a fallen, rotting tree and inoculate the wood. On the ground, a squirrel might defecate on a disturbed area of the forest floor, where a pellet could land near a tree's feeder rootlet that may become inoculated with the mycorrhizal fungus when spores germinate. Alternatively, the red-backed vole and deer mouse might defecate in their subterranean burrows, where fungal spores could come in contact with tree feeder rootlets. If environmental conditions are suitable and root tips are available for colonization, a new fungal colony may be established. Otherwise, hyphae of germinated spores may fuse with an existing nonreproductive part of the fungus and thereby contribute and share new genetic material.

To continue, the transfer of energy captured from the sun by a tree's crown through photosynthesis is transmitted down through the trunk into the roots and their fungal symbiont where—through these same fungi—nutrients and water are conducted from the soil into the roots and through them up the trunk into the crown, perhaps of the squirrel's own nest tree.[22] Although incredibly complicated, this is the tiniest glimpse into the total complexity of complementary functions within a forest's infrastructure.

The more forests are altered by human actions, the more evident becomes the need to understand the interaction of all organisms in the ecosystem. How each component functions is often far more complex than might be anticipated, and the role it plays may be essential in maintaining ecosystem health. The same is true within a city. With respect to the living conduits, neither a forest nor a city could exist without them.

Conduits in Everycity

Although a city is full of conduits, most are either out of sight or virtually unnoticed because they are so commonplace. The conduits we see, but take for granted, are things like power lines that transmit electrical impulses; telephone lines that transmit the human voice; television cables that transmit images; garden hoses and drinking fountains that transmit the flow of water; sidewalks that transmit the flow of pedestrians; bike lanes to transmit the flow of cyclists; streets, county roads, and super highways that transmit the flow of traffic; gutters on houses and along streets, downspouts, storm drains, ditches, and canals that transmit the flow of water; and so on. And, that is just above ground.

Below ground are cables that transmit electricity, human voices, and images as well as pipes that transmit water, human waste, natural gas, oil, and in some places steam. Then, there is the entire subway system to move people from here to there and back again. There are even tunnels beneath rivers, such as the Holland Tunnel for automobiles under the Hudson River, which connects the island of Manhattan in New York City with Jersey City in New Jersey at Interstate 78 on the mainland. And, there is the 31-mile-long

Channel Tunnel for trains from Kent in England under the English Channel to Normandy in France. Completed in 1994, 112 million passengers used the tunnel trains in the first six years of its operation.[23] And, I even remember the trans-atlantic telephone cable on the floor of the ocean, which was revered prior to the invention of communication satellites and computers.

Pipes, highways, railroad tracks, and shipping lanes that bring water, food, and other materials from afar into a city are analogous to fungal mycelia collecting water and elemental nutrients in the soil of a forest and redistributing them to the trees. In turn, sewer systems and septic tanks flow to the call of gravity and reallocate human waste that becomes the food for organisms that form different strands in Nature's food web, a function also found in forests with animal wastes.

Finally, we humans are a large conduit composed of myriad conduits of various sizes for the interchange and passage of food, blood, bile, gas, reproductive, and waste materials—all forms of energy. The same is true for all the other city animals, to say nothing of the conduits embodied in a city's vegetation. Moreover, these living conduits are critical because, whether human, pet, or domestic plant, their life necessities of food, water, and shelter drive the economy in a measurable way.

If you wonder why this is so, keep track of everything you use for a week (for yourself, family, friends, pets, and plants), and if you ponder their origin, you will find a spidery web of interactions that likely covers more of the globe than you would ever imagine. In this way, you are stimulating the global economy by activating a vast network of conduits, be they pipes with water, oil, and natural gas or trade routes on land, on the sea, and in the air. There are still some lessons we have yet to learn from "conduit 101."

A Lesson Ignored in Conduit 101

Conduits transport things from one place to another as well as concentrate the energy of that which they transport. By this I mean that, given a constant volume of water, the force of it coming out of a hose increases (up to some point) with the decreasing diameter of the hose. Anyone who has ever turned a nozzle knows this as a technique used in "power washing" a sidewalk or the outside of a house prior to painting it.

The same dynamic applies to a broad river being suddenly confronted with a narrow canyon through which it must flow. As the volume is constrained, the velocity increases. Here, the positive lesson from the restricting action of a conduit is used in narrowing a stream or river with a concrete structure to increase the water's velocity and therefore its ability to turn turbines and convert the flowing power of the water in electrical power for use by humanity.

I remember seeing a negative dynamic of constraining the volume of water and increasing its velocity at work in a mountain village in eastern Slovakia, where the people had lined the stream flowing through their village with concrete because recent clear-cut logging had denuded hillsides and was causing frequent flooding where heretofore the stream had rarely done so. The concrete confined the water in a narrower, smooth channel as opposed to the one of cobbles and boulders in years past. The consequence of this action was far greater erosion of the stream's channel where the concrete ended than would otherwise have been the case. Moreover, the deleterious effects of such constraint was obvious—the stream's velocity was simply exported downstream, beyond the artificial conduit, where it did much unnecessary damage.

Conversely, a floodplain, as the name implies, is a relatively open expanse that frequently floods. These are areas where storm-swollen streams and rivers spread, decentralizing the velocity of their flow by encountering friction caused by the increased surface area of their temporary bottoms, where much of the floodwater's energy is dissipated. I saw this phenomenon in 1967 while working in the Terai of Nepal, where the Trisuli River had a floodplain more than a mile (1.6 kilometers) wide. I watched the floodplain fill within two days during the monsoon rains.

Although this principle of friction slowing velocity is well known to humanity and is the idea behind disk brakes on automobiles and hand brakes on bicycles and wheelchairs, as well as the design of the parachute, people refuse to accept it whenever they covet the land claimed seasonally by a river. But, when we refuse to accept a lesson from Nature's biophysical principles, it forms an intellectual blockage in our potential understanding of those principles that ultimately govern social-environmental sustainability. The inertia of "informed denial" is the usual procedure when the immediate economic cost of rectifying a mistake is thought to be great, which is nothing less than passing the debt to some other generation. Informed denial, as a remedy, is based on the same level of consciousness of cause and effect that initiated the problem in the first place—a *reaction*, instead of a *response*, that can only compound the problem.

I do not know if what I am about to relate is a unique problem with the U.S. Army Corps of Engineers, but whether it is or not, the refusal to accept the biophysical constraints of Nature is a global issue. The U.S. Army Corps of Engineers is one of the oldest, largest, and most unusual agencies in the federal government. As a bureaucracy of the executive branch that takes its marching orders from Congress, it is a military organization with an overwhelmingly civilian workforce. It is also an environmental regulator that is generally viewed as suspect by environmentalists because members of Congress often authorize projects to steer federal money to their districts, and the corps frequently justifies them with questionable cost-benefit analyses and in the process has reconfigured—artificially engineered—the American landscape. After all, the motto of the corps is "Essayons," French for "let us try." The motto indicates that, throughout its history, the corps has seen Nature as an enemy to subjugate by equating engineering and control with social progress.

For instance, the Corps of Engineers not only designs and builds structures to control flooding and to improve navigation but also issues permits for the alteration of such bodies of water as streams, rivers, marshlands, and estuaries. Although the corps leaders today speak of "working in harmony with nature," the corps still proudly mobilizes its "Annual Campaign Against the Mighty Mississippi." This ongoing battle caused Burton Kemp, a former corps geologist in Mississippi, to say it is not surprising when the corps takes a militaristic approach to the environment. "I'm afraid it's not a Corps of Scientists. It's not a Corps of Biologists," he said with a sigh. "It's a Corps of Engineers."[24]

One aspect of the program conducted by the Army Corps of Engineers is the practice of "channel improvement" in streams and rivers. These improvements lead to a straightening of a stream's channel and a change of its shape that in turn destabilizes the channel. Destabilization of the channel causes downstream effects—erosion of the banks, alterations of the channel's bed, degradation of the aesthetics, and frequently undesirable changes in the composition of the plants and animals that inhabit the stream. From these numerous improvements, each planned on its own isolated rationale (product-oriented thinking), comes the next, larger-order magnitude of massive flooding when the constrained waters release their pent-up energy through a breech in the levees designed to imprison their flow.[25]

To better understand what is meant by the sudden release of pent-up energy, let us go back to the floods in California during the winter of 1996–1997, where sixteen crews worked to shore up some of the 1,100 miles of levees in the Sacramento-San Joaquin Delta. In other areas, crews used sandbags and plastic sheets to shore up critical sections in the 6,000-mile network of levees in northern California.[26]

A dozen major breaks in the levees occurred along the San Joaquin, Mokelumne, and Consumnes rivers, and many other places were threatened because of the pressure of the water day after day. "It's a race with Mother Nature, but right now we're ahead," said Captain Mark Bisbee of the state forestry department.[27]

A levee failed on the San Joaquin River on January 10, 1997, sending work crews fleeing through dense fog as a ninety-foot gap opened, and water began rushing into neighboring fields, where it swamped up to five thousand acres. A break in a levee near Lathrop allowed the flooding of more than twenty-five square miles and damaged as many as four hundred homes.[28]

Water from a ruptured levee on the Feather River, one hundred miles northeast of San Francisco, flooded a farm. The farmer and his wife lost $300,000 worth of cattle. "The sheriff's department just wouldn't let us in," said the farmer, "so 200 head died a slow death. It was gruesome."[29] Some of the farmer's cattle were ensnared in ditches or fences; one cow, snagged on a small gate, had to be burned free with torches. The stench of the rotting animals was everywhere. Thus, floods gave way to fields of death across northern California, where hundreds of drowned cows, horses, and other farm animals—their bloated carcasses tangled in barbed wire or mired in ditches—lay strewn across the soggy landscape.

Although major losses of farm machinery, barns, homes, and wells appeared to be the immediate headache for farmers (including the loss of livestock for some), others were concerned about the survival of their crops. While an estimated 150,000 acres of red winter wheat was likely damaged, wheat farmers were most worried about the potential loss of topsoil to erosion from running water.

With at least three reservoirs nearing their capacities, California water officials said the danger from flooding was far from over as runoff from rain and melting snow continued to build. Because of water released from dams, some rivers (including the Central Valley's Stanislaus and San Joaquin) would continue bulging with runoff inside their eroding levees until February, well after the storm event had passed. "It's unfortunate," said Jeff Cohen, of the California Water Resources Department, "that there is damage downstream, but it's the requirement [draining water from the reservoirs]." He added that the corps requires water to be drained from reservoirs before they reach capacity.[30]

You may be wondering where the people's responsibility lies in this story about flooding. It has to do with priority and people's choices. In bygone times, people lived in California's Central Valley and farmed the land in concert with the rivers, including their periodic floods, because they knew where the floodplain was and respected it. Despite the wisdom exhibited in time past, it has long been American tradition to wrest every useable acre from nature lest an acre be thought of as "unproductive."

If the rivers could be controlled and the flooding stopped, then more of the unproductive acres could be made to produce that which Americans thought desirable. So, dams and levees were constructed. If they failed to produce the desired control, more dams and levees were built. In the end, however, they are proving no match for Nature, as the above story illustrates. But, there is something left unsaid by the story, namely, the choices people made that brought all this about.

Let us look at just four possible choices people *could have made* prior to the floods of 1996–1997: (1) do not live or farm in the floodplain; (2) live and farm in the floodplain *without* dams and levees and plan for, be prepared for, and accept the risk of periodic flooding; (3) live and farm in the floodplain *with* dams and levees in place, thinking the problem of flooding is solved but move after a levee breaks once or twice; (4) live and farm defiantly in the floodplain *with* dams and levees in place regardless of the dire consequences of periodic flooding.

All these choices represent different levels of self-imposed constraints on one's behavior based on different perceived values for monetary gain and lifestyle. The choice that seems to have been generally accepted over time is the last one: live and farm defiantly in the floodplain regardless of consequences. This, then, becomes the primary social constraint or "fixed point" around which all human residential, rural, and commercial development is made to revolve, despite the fact that sooner or later the rivers will remember their floodplains and reclaim them—at least temporarily and at tremendous financial cost.

When their irresponsible risk-taking fails, people want the government (hence society at large) to rescue them, despite their having known the ill-advised risk of building in the floodplain with its inevitable consequences. But, why should the people at large, through personal taxes paid to the government, be expected to bail out those individuals who make unwise choices when they gamble for such high stakes? Do we, through our taxes, bail someone out of financial trouble when they lose heavily in a high-stakes game of craps in a casino in Las Vegas? Building in a floodplain and wagering in a game of craps are both gambling, so what is the difference? Where is personal responsibility? How is one to learn responsibility if one does not have to accept the full measure of the consequences of their choices?

It was not always this way, according to Scott Faber, director of floodplain programs for American Rivers, based in Washington, D.C. According to Faber, "Floods may be acts of God, but flood losses are acts of hubris."[31] Predictable, natural events, like floods, have turned into natural "disasters" that people try to control with dams and levees because housing and commercial development have "flooded" the floodplains.

"At the turn of the century," wrote Faber, "there was virtually no development in floodplains. Over the last 60 years, government programs have assumed responsibility for flood 'control' by building and repairing levees, providing relief, and subsidizing flood insurance. These programs actually put people in harm's way by eliminating incentives for local communities to direct new development away from flood-prone areas."[32]

He went on to say that levees and dams create a false sense of security that encourages people to build in flood-prone areas and thus increases the potential for catastrophe when a levee inevitably fails. Thus, thousands of flood-weary Midwesterners decided in 1993 to stop "playing chicken" with the Mississippi and Missouri rivers. They opted instead for a voluntary program that relocated more than eight thousand homes and businesses, even whole towns, onto the bluffs, so that thousands of people were literally high and dry when floodwaters returned in 1995.

As development continually encroaches on floodplains, said Faber, the rainfall that once was absorbed slowly and naturally by the land now courses rapidly into rivers, which get higher and faster as they flow toward centers of human population. An isolated decision to drain a wetland, till a farm, pave a parking lot, or put in a new street has little measurable effect on flooding by itself, but when combined with thousands of other seemingly unrelated decisions as a cumulative effect, the results can be devastating.

Rather than work together to solve regional problems (systems thinking), Faber says most communities and rural landowners simply pass the water downstream as fast as possible.[33] More dams and levees cannot eliminate human problems associated with flooding because dams and levees fostered the problem in the first place, and more of the same is hardly the cure. Periodic flooding, at times of mammoth proportions, is one of Nature's nonnegotiable constraints, especially during cool, wet periods in the weather cycle, which heretofore have been another of Nature's nonnegotiable constraints. Hence, a wise community will both recognize and bow to Nature when Nature is clearly beyond human control.

Government agencies, in the western United States at least, have not been interested in the long-term future of the landscape because the computed, but often unrealistic, cost-benefit ratio is on the side of utility.[34] Nonetheless, for a system to be viable through some scale of time (be it a forest or a city), its processes must remain functional. A case in point is the service connection.

The Service Connection

All services are ultimately about energy. While energy can be neither created nor canceled, it can be concentrated. "If you follow the energy," wrote author Richard Manning, "you end up in a field somewhere." Since humanity has probably run out of good, arable land, "food is oil." At issue is the green plant's unique ability to convert sunlight into stored energy in the form of carbohydrates through photosynthesis. It is estimated that we humans consume about 40 percent of the Earth's "primary productivity" or the total amount of plant biomass produced each year. Two-thirds of the primary productivity we consume is made up of energy stored in the seeds of three grasses: rice, wheat, and corn.[35]

What happens when the service of the soil that grows the plants as they gather and store the sun's energy is destroyed because humanity abuses the land? Plato (427–347 BCE) recorded it well over two thousand years ago:

> What now remains of the formerly rich land is like the skeleton of a sick man. … Formerly, many of the mountains were arable. The plains that were full of rich soil are now marshes. Hills that were once covered with forests and produced abundant pasture now produce only food for bees. Once the land was enriched by yearly rains, which were not lost, as they are now, by flowing from the bare land into the sea. The soil was deep; it absorbed and kept the water in loamy soil, and the water that soaked into the hills fed springs and running streams everywhere. Now the abandoned shrines at spots where formerly there were springs attest that our description of the land is true.[36]

In the context of this section, *service* refers both to supplies and suppliers from two disparate yet complementary systems: a forest and a city. Suppliers in a city provide such necessities as water, electricity, natural gas, gasoline, and groceries (e.g., rice, wheat, and corn), whereas Nature supplies the urban suppliers with the water, electricity, natural gas, oil for gasoline, and raw produce for groceries (e.g., rice, wheat, and corn). So, how do these disparate yet complementary systems work?

Services in Everyforest and Beyond

Both the quality of human life and the health of our human economy depend on the services performed "free of charge" by Nature, services that are worth many trillions of dollars annually. After all, the bee fertilizes the flower from which it steals. Economic activities that destroy habitats and impair services performed by Nature will create costs to humanity over the long term that undoubtedly will exceed in great measure the perceived short-term economic profits forgone. Most of these services, and the benefits they provide, are part of the commons—everyone's birthright, such as clean water and air. As part of the commons, these services are not traded in economic markets and thus carry no visible price tags that would alert society to their real value, changes in their supply, or deterioration of the underlying biophysical systems that generate them.

These ecological costs are usually hidden from traditional economic accounting but are nevertheless real and borne by society at large—especially, and increasingly, the generations of children. Tragically, a short-term economic focus on current decisions concerning the use of land often sets in motion great costs that, again, are bequeathed by myopic adults to their own children as well as all those of the future. Unfortunately, humanity, as history shows, finds the real value of something only when that something is lost. Put differently, the real value of common things is too often realized only in hindsight.

This said, we can no longer assume that Nature's services will always be available and "free" for our use. Despite our best efforts to think ahead, we will rarely, if ever, be able to ascertain the full impact of our actions and their unforeseen and unpredictable consequences with respect to Nature. The loss of species and habitats and the degradation and simplification of ecosystems both impair Nature's ability to provide the necessary services we depend on for life. Losses, by definition, are irreversible and irreplaceable.

The inherent services performed by Nature constitute the invisible foundation that is at once the wealth of every human community and its society as well as the supporting basis of our economies. For example, we rely on oceans to supply fish; forests to supply water, wood, and new medicines; streams and rivers to transport the water from its source to a point where we can access it; soil to grow food; grasslands to raise livestock, and so on.

Although we base our livelihoods on the expectation that Nature will provide these services indefinitely and free of charge, the economic system to which we commit our unquestioning loyalty undervalues, discounts, or ignores these services when computing the gross domestic product. Nature's services are thus measured poorly or not at all.

Because of the importance of Nature's inherent services, usually thought of as ecosystem functions, it is worthwhile to examine one in greater detail—pollination. Wild and semiwild pollinators are responsible for the fertilization of 80 percent of all cultivated crops (1,330 varieties, including fruits, vegetables, coffee, and tea). Between 120,000 and 200,000 species of animals perform this service.

Bees are enormously valuable to the functioning of virtually all terrestrial ecosystems and such worldwide industries as agriculture. Pollination by European honeybees, for example, is sixty to a hundred times more valuable economically than is the honey they produce. In fact, the value of wild blueberry bees is so great that farmers who raise blueberries refer to them as "flying $50 bills."[37]

While more than half of the honeybee colonies in the United States have been lost within the last 50 some years, 25 percent were lost between 1994 and 1999. Widespread threats to bees and other pollinators are fragmentation of and outright destruction of habitat (hollow trees for colonies in the case of wild honeybees), intense exposure to pesticides, a generalized loss of nectar plants to herbicides, as well as the gradual deterioration of "nectar corridors" that provide sources of food to migrating pollinators.

In Germany, for instance, the people are so efficient at weeding their gardens that the nation's free-flying population of honeybees declined rapidly, according to Werner Muehlen of the Westphalia-Lippe Agricultural Office. Bee populations shrank by 23 percent across Germany over the decade of the 1990s, and wild honeybees are all but extinct in Central Europe. To save the bees, said Muehlen, "Gardeners and farmers should leave at least a strip of weeds and wildflowers along the perimeter of their fields and properties to give bees a fighting chance in our increasingly pruned and … [sterile] world."[38]

Besides an increasing lack of food, one-fifth of all the losses of honeybees in the United States are from exposure to pesticides. Wild pollinators are even more vulnerable to pesticides because, unlike hives of domestic honeybees that can be picked up and moved prior to the application of a chemical spray, colonies of wild pollinators cannot be purposefully relocated. On the other hand, because domesticated honeybees service only 15 percent of the world's major crops, they cannot be expected to fill the gap by themselves if wild pollinators are lost.

Ironically, economic valuation of products, as measured by the gross domestic product, actually fosters many of the practices employed in modern intensive agriculture and modern intensive forestry that literally curtail the productivity of crops by reducing pollination. An example is the high level of pesticides used on cotton crops to kill bees and other insects that reduce the annual yield in the United States by an estimated 20 percent or

$400 million. In addition, herbicides often kill the plants needed by pollinators to sustain themselves when not fertilizing crops. Finally, the practice of squeezing every last penny out of a piece of ground by plowing the edges of fields to maximize the planting area can reduce yields by removing nesting and rearing habitats for pollinators.

Unfortunately, too many people are fueled by their unquestioning acceptance of current economic theory, even as it designs, condones, and actively encourages the above-mentioned destructive practices. Such people simply *assume* that the greatest value one can derive from an ecosystem, such as a forest, is that of maximizing its productive capacity for a single commodity to the exclusion of all else. Single-commodity production, however, is usually the least-profitable and least-sustainable way to use a forest because it simply cannot compete with the enormous value of nontimber services, such as producing oxygen, capturing and storing water, holding fertile soils in place, and maintaining habitat for organisms that are beneficial to the economic interests of people. These are forgone when the drive is to maximize a chosen commodity to bolster short-term monetary profits.[39]

Not surprisingly, the undervalued, discounted, or ignored uses of the forest are more valuable and more sustainable and benefit a far greater number of people in the long term than does the production of wood fiber in the short term. We need to remember that Nature supplies the services, commodities, and amenities to the suppliers who supply services to the citizen of the city.

Services in Everycity

Services within a city are ultimately supplied by Nature from the "raw materials" humanity finds and uses. That notwithstanding, what is seldom understood in a city is that the quality and quantity of humanity's crucial supplies, such as clean air to breathe, clean water to drink, and fertile soil for growing healthy food, are dependent on the consciousness with which humanity abides by the biophysical principles that govern Nature. In this sense, the services a city requires and obtains from Nature are part and parcel of the natural capital, the real wealth that sustains us.

Take water, for instance. Every plant, pet, and person in the city requires water, be it for drinking; cooking; washing dishes, windows, and clothes; flushing toilets; painting with watercolors; brushing teeth; taking a shower or bath; use by dentists, doctors, veterinarians, paper manufacturers; and on and on and on. Why, therefore, do not citizens insist that government at all levels, as well as the timber industry, livestock industry, and mining industry, take the best possible care of the nation's water catchments from which every city garners its supply of water?

City dwellers frequently seem oblivious to the fact that all we do, with rare exception, is serve one another in some way. People speak about the "service industry" as though it is somehow apart from all the other facets of normal human endeavors in a city. True, not all people are kind, not all people serve graciously, some people gouge their customers in pricing their goods, and some people are just downright rude and dishonest in their dealings with others. But in the main, it seems most people are serving one another with at least a modicum of kindness. Of course, service providers can treat their customers well, behavior that generally leads to customers returning again and again to the service providers who treated them the best.

If we all treat one another with the best principles of human relationships, it is analogous to complying with Nature's biophysical principles by taking responsibility for our own behavior. In other words, if I want to become acquainted with you, it is incumbent on me to determine how I must treat you in order to encourage you to reciprocate in kind. Thus, for me to receive the best service, it is my responsibility to initiate a good relationship with the person serving me. Likewise, to have an adequate supply of good-quality resources in the form of services from Nature to run our cities, we are obliged to care for the land in a way that perpetuates the natural capital we, and all generations of the future, require for a good-quality life. Here, the bottom line is that, by treating one another, the land, and Nature with respect, we are uniting the two disparate systems into a single, self-reinforcing feedback loop of reciprocity that can be perpetrated through time.

But, to bring this about, we need to view one another and ourselves differently, a necessity that requires a brief, generalized visit to the hunter-gatherers of antiquity. If you are wondering why we need to visit them, the answer is simple: to understand what we have forgotten—how to live in harmony with one another and our environment.

Who Benefits, Who Pays?

The hunting-gathering peoples of the world—Bedouins, Australian aborigines, African Bushmen, and similar groups—represent the oldest and perhaps the most successfully adapted human beings. Virtually all of humanity lived by hunting and gathering before about twelve thousand years ago. Hunters and gatherers represent the opposite pole of the densely packed, harried urban life most people of today experience. Yet, the life philosophy of those same hunter-gatherers may hold the answer to a central question plaguing humanity in the twenty-first century: Can people live harmoniously with one another and Nature?

Until 1500 AD, hunter-gatherers occupied fully one-third of the world, including all of Australia, most of North America, and large tracts of land in South America, Africa, and northeast Asia, where they lived in small groups without the overarching, disciplinary umbrella of a state or other centralized authority. They lived without standing armies or bureaucratic systems, and they exchanged goods and services without recourse to economic markets or taxation.

With relatively simple technology, such as wood, bones, stones, fibers, and fire, they were able to meet their material requirements with a modest expenditure of energy and had the time to enjoy what they possessed materially, socially, and spiritually. Although their material wants may have been few and finite and their technical skills relatively simple and unchanging, their technology was, on the whole, adequate to fulfill their needs, a circumstance that says the hunting-gathering peoples were the original affluent societies. Clearly, they were free of the industrial shackles in which we find ourselves as prisoners at hard labor caught seemingly forever between the perpetual disparity of unlimited wants and insufficient means.

Evidence indicates that these peoples lived surprisingly well together despite the lack of a rigid social structure, solving their problems among themselves, largely without courts and without a particular propensity for violence. They also demonstrated a remarkable ability to thrive for long periods, sometimes thousands of years, in harmony with their environment. They were environmentally and socially harmonious and thus sustainable because they were egalitarian, and they were egalitarian because they were socially and environmentally harmonious. They intuitively understood the reciprocal, indissoluble connection between their social life and the sustainability of their environment.

Sharing was the core value of social interaction among hunter-gatherers, with a strong emphasis on the importance of generalized reciprocity, meaning the unconditional giving of something without any expectation of immediate return. The combination of generalized reciprocity and an absence of private ownership of land has led many anthropologists to depict the hunter-gatherer way of life as a "primitive communism," in the true sense of communism, wherein both prosperity and property are owned in common by all members of a classless community or society.

Even today, there are no possessive pronouns in aboriginal languages. The people's personal identity is defined by what they give to the community.

> "I am because we are, and since we are, therefore I am" is a good example of the "self-in-community" foundation that gives rise to a saying in Zulu, such as *umuntu ngumuntu ngabantu*—"It is through others that one attains selfhood."[40]

Hunter-gatherer peoples lived with few material possessions for hundreds of thousands of years and enjoyed lives that were in many ways richer, freer, and more fulfilling than ours. These nomadic peoples were (and are) economical in every aspect of their lives, except in telling stories. Stories passed the time during travel, archived the people's history, and passed it forward as the children's cultural inheritance.[41]

These peoples so structured their lives that they wanted little, required little, and found what they needed at their disposal in their immediate surroundings. They were comfortable precisely because they achieved a balance between what they needed and wanted by being satisfied with little. There are, after all, two ways to wealth—working harder or wanting less.

The !Kung Bushmen of southern Africa, for example, spent only twelve to nineteen hours a week getting food because their work was social and cooperative, so they obtained their particular food items with the least possible expenditure of energy. Thus, they had abundant time for eating, drinking, playing, and general socializing. In addition, young people were not expected to work until well into their twenties, and no one was expected to work after age forty or so.

Hunter-gatherers also had much personal freedom. There were, among the !Kung Bushmen and the Hadza of Tanzania, for instance, either no leaders or only temporary leaders with severely limited authority. These societies had personal equality in that *everyone* belonged to the same social class *and* had gender equality. Their technologies and social systems, including their economies of having enough or a sense of "enoughness," allowed them to live sustainably for tens of thousands of years. One of the reasons they were sustainable was their lack of connection between what an individual produced and that person's economic security, so acquisition of things to ensure personal survival and material comfort was not an issue.[42]

In the beginning, nomadic hunters and gatherers, who have represented humanity for most of its existence, probably saw the world simply as "habitat" that fulfilled all of their life's requirements, a view that allowed the people to understand themselves as part of a seamless community.

With the advent of herding, agriculture, and progressive settlement, however, humanity created the concept of "wilderness," and so the distinctions between "tame" (meaning "controlled") and "wild" (connoting "uncontrolled") plants and animals began to emerge in the human psyche. Along with the notion of tame and wild plants and animals came the perceived need not only to "control" space but also to "own" it through boundaries in the form of corrals, pastures, fields, and villages. In this way, the uncontrolled land or wilderness of the hunter-gatherers came to be viewed in the minds of settled folk as "unproductive," "free" for the taking, or as a threat to their existence.

So it is that agriculture brought with it both a sedentary way of life and a permanent change in the flow of living. Whereas the daily life of a hunter-gatherer was a seamless whole, a farmer's life became divided into "home" and "work." While a hunter-gatherer had intrinsic value as a human being with respect to the community, a farmer's sense of self-worth became extrinsic, both personally and with respect to the community as symbolized by, and permanently attached to, "productivity"—a measure based primarily on how hard a person worked and thus the quantity of goods or services produced.

In addition, the sedentary life of a farmer changed the notion of "property." To the hunter-gatherers, mobile property, that which one could carry with them (such as one's hunting knife or gathering basket) could be owned, but fixed property (such as land) was to be shared equally through rights of use but could not be personally owned to the exclusion of others or the detriment of future generations. This was such an important concept that it eventually had a word of special coinage, *usufruct*.

According to the 1999 *Random House Webster's Unabridged Dictionary*, usufruct is a noun in Roman and civil law. Usufruct means that one has the right to enjoy all the advantages derivable from the use of something that belongs to another person provided the substance of the thing being used is neither destroyed nor injured.

The dawn of agriculture ultimately gave birth to civilizations and created powerful, albeit unconscious, biases in the human psyche. For the first time, humans clearly saw themselves as distinct from and—in their reasoning at least—superior to the rest of Nature. They thus began to envision themselves as masters of, but no longer as members of, Nature's biophysical community of life.

To people who lived a sedentary life, like farmers, land was a commodity to be bought, owned, and sold. Thus, when hunter-gatherer cultures, such as the American Indians, "sold" their land to the invaders (in this case Europeans), they were really selling the right to "use" their land, not to "own" it outright as fixed property, something the Europeans did not understand. The European's difficulty in comprehending the difference probably arose because once a sedentary lifestyle is embraced, it is almost impossible to return to a nomadic way of life, especially the thinking that accompanies it.

We, as individuals, may despair when we contemplate the failure of so many earlier human societies to recognize their pending environmental problems as well as their failure to resolve them—especially when we see our local, national, and global society committing the same kinds of mistakes on an even larger scale and faster time track. But, the current environmental crisis is much more complex than earlier ones because modern society is qualitatively different from previous kinds of human communities. Old problems are occurring in new contexts, and new problems are being created, both as short-term solutions to old problems and as fundamentally new concepts. Pollution of the world's oceans, depletion of the ozone layer, production of enormous numbers and amounts of untested chemical compounds that find their way into the environment, and the potential human exacerbation of global climate change were simply not issues in times past. But, they are the issues of today.

There are lessons we, as a society today, can *relearn* from the people who lived, and the few who still live, a hunter-gatherer way of life. I say, "relearn" because, as writer Carlo Levi once said, "the future has an ancient heart."[43]

- Life's experiences are personal and intimate.
- Sharing life's experiences by working together and taking care of one another along the way is the recipe for and price of sustainability.

- Cooperation and coordination, when coupled with sharing and caring, preclude the perceived need to compete, except in play—and perhaps in storytelling.
- The art of living lies in how we practice relationship—beginning with ourselves—because practicing relationship is all we humans ever do in life.
- Leisure is affording the necessary amount of time to fully engage each thought we have, each decision we make, each task we perform, and each person with whom we converse in order to fulfill the relationship's total capacity for a quality experience.
- Simplicity in living and dying depends on and seeks things small, sublime, and sustainable.
- There is more beauty and peace in the world than ugliness and cruelty.
- Any fool can complicate life, but it requires a certain genius to make and keep things simple.
- For a group of people to be socially functional, they must be equally informed about what is going on within the group; in other words, there must be no secrets that are actually or potentially detrimental to any member of the group.
- Separating work from social life is not necessary for economic production—and may even be a serious social mistake.
- By consciously limiting our "wants," we can have enough to comfortably fulfill our necessities as well as some of our most ardent desires—and leave more for other people to do the same.
- Simplicity is the key to contentment, adaptability, and survival as a culture; beyond some point, complexity becomes a decided disadvantage with respect to cultural longevity, just as it is to the evolutionary longevity of a species.
- The notion of scarcity is largely an economic construct to foster consumerism and increase profits but is not necessarily an inherent part of human nature. (We need to overcome our fear of economically contrived scarcity and marvel instead at the incredible abundance and resilience of Earth.)
- Linking individual well-being strictly to individual production is the road to competition, which in turn leads inevitably to social inequality, poverty, and environmental degradation.
- Self-centeredness and acquisitiveness are not inherent traits of our species but rather acquired traits based on a sense of fear and insecurity within our social setting that fosters the perceived need for individual and collective competition, expressed as the need to impress others.

- Inequality based on gender or social class indicates a behavior based on *fear* disguised as "social privilege."
- Mobile property, that which one can carry with them, can be owned, whereas fixed property—such as land that may be borrowed—is to be shared equally through rights of generational use but cannot be personally owned to the detriment of any generation.
- Placing material wealth, as symbolized by the money chase, above spirituality, Nature, and human well-being is the road to social impoverishment, environmental degradation, and the collapse of societies and their life-support systems.

So, where are we today? We are the exact antithesis of the hunter-gatherers in many respects, such as who we are, how we obtain resources, what we own, our connection with Nature, and who benefits and who pays. I am, however, going to focus on our connection with Nature because, in a sense, the others are embodied in the characteristics of that relationship.

The hunter-gatherers knew themselves to be an inseparable part of Nature and so did their best to honor Nature by blending in with the seasonal cycles of birth and death, of hunter, gatherer, and hunted. Through their spirituality and myths, they sought to understand the "Nature gods," appease them, and serve them so they might continue to be generous in the future.

We city folks, on the other hand, have all but lost our conscious connection with Nature, in part because a number of modern religions, such as Christianity and Judaism, deem humanity to be separate from and above all other life on Planet Earth. In addition, we live in protective "boxes" of one sort or another into which our daily necessities are transported—including our experience of the outer world via television. Consequently, we rarely experience the night sky, the seasonal flights of migrating geese, or the wide-open spaces that are as yet uncluttered by the trappings of humanity. And, those city folks who do hunt normally do so with high-powered rifles that make their game into the abstractions of sport and trophies—not lives taken with reverence for the necessity of food.

The hunter-gatherers lived lightly on the land, honoring its cycles, being patient with Nature's pace, taking only what they needed and thereby allowing the land to renew itself before they took from it again. In this way, generations passed through the millennia, each tending to be at least as well off as the preceding one.

Because we have a propensity to see Nature as a commodity to be exploited for our immediate benefit, we are, at best, short-changing the generations of the future by passing forward unpaid environmental bills and, at worst, blatantly stealing their inheritance and thus setting all generations on a course toward environmental bankruptcy. The first is irresponsible, the latter unconscionable.

Moreover, while the hunter-gatherers lived an effective life, we are focused almost totally on efficiency. But, they are *not* the same thing.

The Transportation System: Efficiency or Effectiveness?

First, a brief lesson is needed to distinguish the difference between *efficiency* and *effectiveness*. Pine trees cast on the winds of fortune a prodigious amount of pollen to be blown hither and yon. I say "the winds of fortune" because it takes an inordinate amount of pollen riding the vagaries of air currents to come in contact with and fertilize enough pine seeds to keep the species viable through time. Although an extremely *inefficient* mode of pollination in that many, many more grains of pollen are produced than are used to fertilize the available pine seeds, the system is highly *effective*, as evidenced by the persistence of pine trees through the ages. And, if you are wondering what happens to all the "unneeded" grains of pollen, they are eaten by a variety of organisms that benefit from an extremely rich source of nutriment. Nothing in Nature is wasted. "Waste," as people think of it, is an *economic concept*—not an ecological one.

If you are wondering what pollen grains have to do with a system of transportation, the answer is simple. They have to do with the *focus* of the transportation system—connectivity of the landscape (based on social-environmental effectiveness and thus sustainability) or fragmentation of a landscape (based on economic efficiency).

Transportation in Everyforest

To date, the vast majority of the transportation systems connected with extraction of timber are designed and located based on the premise that *efficiency* is the primary concern—least cost, maximum financial gain. The result of such product-oriented thinking in the placement and construction of roads both fragments forest habitat unnecessarily and negatively affects the quality, quantity, and distribution of water in the soil. The construction and use of a road—especially a poorly placed and designed one—severely disturbs the soil, which in turn increases the rate of runoff, hence erosion; reduces the flow of subsurface water; and alters the equilibrium of shallow groundwater.

Speaking of poorly placed roads, protecting bodies of water in a forest for their capacity to store this precious liquid is crucial. To this end, it would be wise to place a *permanent moratorium* on the construction of roads near water. Along with this moratorium, ecological *effectiveness* would dictate abandoning and permanently obliterating as many roads as possible that parallel a stream or river as well as removing all culverts and bridges rendered unnecessary. Further, it would be environmentally prudent to relocate roads deemed necessary to the tops of ridges, even if it increased the length of a given road. These actions alone would do much to reconnect the forest habitat and protect the quality of humanity's supply of potable water and thus vital services in distant cities.

Ridge-top roads are important because, unless water infiltrates deep into the soil of a water catchment, it runs downhill and reaches the cut bank of a logging road or even a major highway, which brings it to the surface, collects it into a ditch, and puts it through a culvert to begin infiltrating again. The water then meets another road cut and so on. Water is sometimes brought to the surface three, four, or more times before reaching a stream.

Water is purified by its journey through the deeper soil but not by flowing over the surface of the ground. Roads bleed water from the soil the same way cuts in the bark bleed latex from a rubber tree or sap from a sugar maple. In fact, ditches and gullies, such as those that form on the downhill side of culverts passing under roads, function as pathways down which water flows. The denser the network of roads, the greater the drainage of water over the soil's surface and the less time it takes for peak flows to occur in streams and thus dissipate the potential storage capacity of the soil.

This poses a question: How deep into the soil is deep enough for water to avoid the ditches at the bases of banks alongside roads? I have seen roadbeds blasted out of solid rock to depths of fifty or sixty feet (fifteen to eighteen meters), and I have seen water seeping out the "bottom" of this same rock into the roadside ditches in July and August, a predicament symptomatic of the disruption in the flow of water. In this way, precious water is brought to the surface of the ground, where it both evaporates and becomes polluted by sediment, oil, and chemicals from the road's surface and human garbage in the ditch. Consequently, roads have a negative cumulative effect on the hydrological cycle of a water catchment and on the purity of the water that ultimately reaches human communities.

Disrupting the flow of water through the soil on steep slopes, even forested slopes, can cause instability and increase erosion during a severe rainstorm or as snow melts. Such conditions in the vicinity of the seeping water can cause soils to become saturated, with little or no infiltration, which in turn weakens them and leads to greater erosion.[44]

Even if water flowing through the soil were a constant, a variable is introduced with construction of a single logging road. What is more, constructing and maintaining multiple logging roads to extract timber only compounds the variable. In addition, intensive forestry, such as clear-cutting, alters the water regime, which affects how the forest grows. In this way, a self-reinforcing feedback loop of biophysical degradation in a water catchment is created, altering the soil-water regime, which in turn alters the sustainability of the forest, which in turn affects the soil-water regime in a never-ending cascade of cause and effect. Eventually, the negative effects are felt in those communities that are dependent on a given water catchment or drainage basin for their supplies of potable water.

In essence, there is nothing—*nothing*—environmentally friendly about a road, any road.[45] Roads act both as conduits for the invasion of exotic plants throughout most of North America and augment the illegal killing of wildlife.[46] Roads serve one purpose, and one purpose only: to allow human access to an area in a motorized vehicle of some kind. A few succinct data points from early 1998 will put the issue of roads, on national forests alone, into perspective: An estimated 433,000-mile spider web of official roads crisscrossed our national forests in addition to some 60,000 miles of "ghost roads" created by the repeated use of four-wheel drive and off-road vehicles that were not officially out there and that did not count as part of the extensive network of county, state, and national roads.[47]

Every forest and city ideally needs open spaces to maximize its health. Although many of today's forested areas are too small when it comes to accommodating open spaces of any real consequence, our national forests are a different story. In this case, roadless areas could serve as "open spaces."

A *roadless area* is a forest with no evidence of previous logging and of sufficient size (500 acres or larger; 202 hectares or larger) and configuration that could maintain ecological integrity. Protecting such areas is critical because, in our burgeoning, product-oriented society, one of the most insidious dangers to indigenous forests—those with minimal, disruptive human intrusion—is the sadly mistaken perception that there is no intrinsic value in maintaining such a forest. By intrinsic value, I mean its value as a blueprint of what a sustainable forest is, how it functions, its educational value, its spiritual value, or any other value that does not turn an immediate, economic profit.

Although some roadless areas have survived in the modern world, they not only are changing constantly but also are in danger of natural, catastrophic disturbance, such as large, intense fires. Consequently, they cannot simply be left alone because, should one burn, it is gone forever. True, the acreage is still in place, the land is still roadless, and the trees will eventually grow back, but the indigenous forest is gone—and that is the point.

Furthermore, roadless areas are at the most natural end of the aforementioned continuum between naturalness and culturalness. As such, they are critical areas in which to learn about the long-term processes and trends of a naturally developing forest. It is thus imperative to emulate the biophysical processes that keep them healthy, developing along their ordained trajectories, and protect them, as much as possible, from catastrophic events so future generations can benefit from their existence. In other words, these roadless areas are far too valuable as biophysically functional blueprints to lose. Accordingly, a moratorium on the disturbance of roadless areas is needed because once they are gone—they are gone, just like open spaces in and around our cities.

Transportation in Everycity

When the system of transportation becomes the centerpiece of a city's development, the city is placing its primacy of human relationship on the efficiency of mass movement from one place to another, which coincidentally determines where and how the population and open spaces will be situated. Here, a fundamental question might be posed: Does building more and more roads really relieve congestion, which after all seems to be what drives the design of a transportation system?

According to Bill Bishop, editorial page columnist for the *Herald-Leader* in Lexington, Kentucky, building more roads only *adds* to congestion.[48] A parallel can be seen in the continual growth of "self-storage units" as they continue springing up across cities, where they fill the role of "supplemental garbage dumps" for all the *stuff* Americans are constantly purchasing but have nowhere in their homes to keep. This compulsive buying is the same as our compulsive road building. The former does not fill the emptiness of our souls, and the latter does not relieve vehicular congestion.

If our cities' roads are congested and we build more roads to relieve the congestion, will we not just fill the new roads again to the point of overflowing? It seems to me one could logically say: *like our houses, so our roads.*

"Trying to pave your way out of traffic congestion," wrote Bishop, "is like trying to eat your way back into your high school jeans. Cars fill in the new pavement, just like middle age created the market for Dockers." Although it seems counterintuitive, said Bishop, building more roads actually leads to more traffic. On the other hand, he continued, closing roads, or even narrowing streets, does not create more congestion—it tends to cut the volume of traffic, especially in cities.

"Lord knows," said Bishop, "the evidence of this phenomenon is stalled in full view of most citizens. As soon as roads are built, they're filled. And to relieve the new traffic, we build new roads. You'd think somebody would connect the dots." What dots? The dots that illustrate the level of consciousness causing a problem in the first place is not the level of consciousness that can solve it. A higher level of consciousness is required—recognizing, accepting, and acting on the evidence under our noses *is* connecting the dots.

Some people have connected the dots, quipped Bishop. "Adding new roadways and widening older ones was seen as the way to solve the problem," observed the Texas Transportation Institute in a study of city traffic. "In most cities, this new roadway capacity was quickly filled with additional traffic, and the old problems of congestion returned."[49]

On the other hand, researchers at the University College of London, England, examined sixty cases from around the world in which roads had been closed. They found that a goodly portion of the traffic that once used the roads simply "evaporated." The cars and trucks were not simply rerouted on nearby streets but disappeared altogether.

On average, one-fifth of the vehicular use, and in some cases as much as 60 percent, went away once a road was closed, and the full volume of vehicles did not reappear once a road was reopened. The Tower Bridge in London, for example, was temporarily closed in 1994, and the traffic dispersed. Three years after the bridge was reopened, traffic still had not returned to its former level.[50]

Writers James O. Wilson and James Howard Kunstler argued in the online magazine *Slate* that "we have transformed the human ecology of America, from sea to shinning sea, into a national automobile slum."[51] Bishop, meanwhile, wondered if we just "can't remember any other way to live?" At this juncture, you might well be wondering what any of the foregoing has to do with the connectivity or the fragmentation of a landscape.[52]

That is a good question because if the transportation system is the pivot around which a city's planning centers, it will hide the night sky with light pollution and disguise the birdsong with noise pollution, simultaneously precluding much of Nature through habitat fragmentation. These alternatives leave a city only two options in planning its transportation system: ecological constraints that place the greatest emphasis on quality of life, both short and long term, or economic constraints that focus the greatest emphasis on maximizing immediate and short-term profit margins.

If a city chooses to design its transportation system around ecological constraints, it will place the components of the system where they will best honor the integrity and connectivity of the available habitat, including the city's interface with its surrounding landscape. When a transportation system is planned around ecological constraints (effectiveness), the probability of being able to have a relatively good system of open spaces is greatly increased.

On the other hand, if a city chooses to design its transportation system around economic constraints (efficiency), available habitat will suffer far greater fragmentation than if an open-space system itself had driven the city's planning and its implementation. Under the efficiency mode, open space, as a viable system, will be forgone because fragmentation of habitat is inevitably maximized, as are noise and light pollution. There is also a greater likelihood that exotic and naturalized species would colonize the remaining landscape because the transportation system—acting as a conduit of immigration—puts ever-more outside pressure on the survival of indigenous species.

Nevertheless, the choice between ecological constraints (with its emphasis on quality of life) or economic constraints (with its emphasis on profit margins) is seldom posed. Peter Headicar, in transport planning at Oxford Brookes University in England, stated this perpetual lack of choice eloquently. He said that basic questions about the urban future in the context of transportation are not often asked because "they are both politically uncomfortable and tractable only over the longer term—hence conveniently forever deferrable in the present."[53]

In fact, building new, bigger, and faster roads has become the major preoccupation of government at all levels in the United States during much of the past sixty or more years. Billions of dollars have been, and continue to be, spent widening and straightening streets and highways in almost every urban and rural setting throughout the United States.

An entirely new profession—traffic engineers—has materialized to accomplish this feat, and they now wield far more influence over the planning and layout of communities than do elected officials, businesspeople, or citizen groups combined. Traffic engineers have a single goal—enable cars and trucks to move faster, easier, and cheaper through both town and countryside, which increasingly affects the hydrological continuum.[54]

Even on a small scale, say a new housing development, roads and streets are paved, creating an impervious coating over the surface of the land. This impervious layer prevents the water, both as rain and melting snow, from infiltrating into the soil, where it can be stored and purified and can recharge existing aquifers and wells. Instead, the water remains on the surface, where it mixes with pollutants that collect on the pavement.

Because paved roads and streets are lined with curbs and gutters, the now-polluted water is channeled into a storm drain. In addition, each house has an impervious roof that collects water and channels it into gutters along the edge of the roof. On collecting water, the gutters channel it, more often than not, out to the street, where it joins water from the street going down the storm drain. It is then conducted either directly into a sewage treatment plant or directly into a ditch, stream, or river.

In any event, the water is not usable by the local people. Beyond that, the storm water either adds to the cost of running the treatment plant, where it must be detoxified, or it pollutes all the waterways through which it flows, from its point of origin into the ocean.

The effect of roads, streets, parking lots, and the area covered by houses, all of which eliminate the infiltration of water, is cumulative. Enough roads, streets, parking lots, and roofs over time can alter the soil-water cycle as it affects a given community. Remember, the quality and quantity of water is a biophysical variable, irrespective of the fact that many product-oriented economists and product-oriented "land developers" deem it an economic constant.

All these effects are hidden for some time in both the invisible present and the ecological lag period wherein they work synergistically in shifting the landscape from the more natural end of the continuum to the more cultural end. Beyond some point, these effects upset the ecological integrity and ultimately affect the quality of life, almost inevitably in the negative, and there is no backup system to protect the quality of life.

The Importance of Backups

As previously mentioned, when an ecosystem changes and is influenced by increasing magnitudes of stresses, the replacement of a stress-sensitive species with a functionally similar but more stress-resistant species preserves the system's overall productivity. Such replacements of species—backups—can result only from within the existing pool of biodiversity.

Backups are a fundamental design of Nature that we humans increasingly ignore in our cities at our peril, and yet backups are an essential part of daily life. Built-in backup systems give ecosystems and cities the resilience to either resist change or bounce back after disturbance. In this way, a backup system acts as an insurance policy that effectively protects the continuance of a system after a major disruption, be it in a city or a forest, despite the fact that some might deem this notion to be an inefficient waste of money.

And yet, with the proven probability that our ever-increasing reliance on electrical power is making us vulnerable to "electrical blackouts," virtually every village and city is converting as much of their functional systems to computerization as possible, thereby eliminating the manual component as unnecessary duplication—*redundancy*—and thus a waste of money. Here, it may be instructive to take a lesson from the forest about the value of backup systems.

Backup Systems in Everyforest

Backup systems, in the biological sense, are comprised of the various functions of different species that act as an environmental insurance policy. To maintain this insurance policy, an ecosystem needs three kinds of diversity: biological, genetic, and functional. *Biological diversity* refers to the richness of species in any given area. *Genetic diversity* is the way species adapt to change. The most important aspect of genetic diversity is that it can act as a buffer against the variability of environmental conditions, particularly in the medium and long terms. And, *functional diversity* equates to the biophysical processes that take place within the area. The upshot is that healthy environments can act as "shock absorbers" in the face of catastrophic disturbance.

To better understand this concept, think of each of these kinds of diversity as the individual leg of an old-fashioned three-legged milking stool. When so viewed, it soon becomes apparent that if one leg (one kind of diversity) is lost, the stool will fall over. Fortunately, however, biological diversity passed forward through genetic diversity effectively maintains functional diversity.

This backup results in a stabilizing effect similar to having a six-legged milking stool, but with two legs of different kinds of wood in each of three locations. So, if one leg is removed, it initially makes no difference which one because the stool will remain standing. If a second leg is removed, its location is crucial because, should it be removed from the same place as the first, the stool will fall. If a third leg is removed, the location is even more crucial because removal of a third leg has now pushed the system to the limits of its stability, and it is courting ecological collapse in terms of the value we, as a society, placed on the system in the first place. The removal of one more piece, no matter how well intentioned, will cause the system to shift dramatically, perhaps to our long-term social detriment.

When, therefore, we humans tinker willy-nilly with an ecosystem's composition and structure to suit our short-term economic desires, we risk losing species, either locally or totally, and so reduce first the ecosystem's biodiversity, then its genetic diversity, and finally its functional diversity in ways we might not even imagine. With decreased diversity, we lose existing choices for influencing our environment. This loss may directly affect our long-term economic viability because the lost biodiversity can so alter an ecosystem that it is rendered incapable of producing that for which we once valued it or that for which we, or the next generation, could potentially value it again. Maintaining backup systems is thus a critical link in the shared relationship between forests and cities.

Backup Systems in Everycity

As its computer network was being constructed, the officials of my hometown, like businesses and communities everywhere, were increasingly focused on all conceivable aspects of *efficiency* in order to eliminate as much perceived "redundancy" (really backup systems) as possible in everyday activities because they were seen as a waste of money. Accordingly, everything in my hometown that could be computerized was computerized to eliminate the *unwanted redundancy* of manual control—everything, that is, except our fortuitously overlooked water supply (or perhaps there was a systems thinker in the proverbial woodpile). Today, should the computer program that controls the water supply suddenly fail, we would still have water because of the unwanted backup system of manual override. The ability to manually override a computer failure (while thought *inefficient*) is *effective* in giving my hometown the resilience to overcome a potentially disastrous circumstance and remain viable, while other, more "efficient" communities may not be so fortunate.

So, it is critical that we humans rethink the wisdom and utility of the economic notion of efficiency versus the functional reality of effectiveness and come to grips with the differences. If we do not, we may well reap the unwanted consequences of a lesson not learned in Nature's class on "sustainable design 101."

Summation

In Chapter 7, we glimpsed a few similarities in the infrastructural maze of a forest and a city. The social-environmental infrastructure of which we are an inseparable part is the interplay among Nature's biophysical principles, humanity's attempt to impose its will in order to create the design of its choice, and the processes through which humanity attempts to control the biophysical principles while exerting its will. In this regard, I discussed five primary things: (1) the energy interchange system, (2) conduits of interchange, (3) the service connection, (4) the transportation system (efficiency or effectiveness), and (5) the importance of backups.

Part III focuses on the shared relationships between Everyforest and Everycity, beginning in Chapter 8 with a discussion of cumulative effects, lag periods, and thresholds.

Notes

1. William S. Pretzer. Technology education and the search for truth, beauty and love. *Journal of Technology Education*, 8(2) (Spring 1997). http://scholar.lib.vt.edu/ejournals/JTE/v8n2/pretzer.jte-v8n2.html (accessed January 20, 2009).
2. The general discussion of mycorrhizae is based on David Read. The ties that bind. *Nature*, 388 (1997):517–518; David Read. Plants on the web. *Nature*, 396 (1998):22–23; Marcel G. A. van der Heijden, John N. Klironomos, Margot Ursic, and others. Mycorrhizal fungal diversity determines plant biodiversity, ecosystem variability and productivity. *Nature*, 396 (1998):69–72; Anna S. Marsh, John A. Arnone, Bernard T. Bormann, and John C. Gordon. The role of *Equisetum* in nutrient cycling in an Alaskan shrub wetland. *Journal of Ecology*, 88 (2000):999–1011; Chris Maser, Andrew W. Claridge and James M. Trappe. *Trees, Truffles, and Beasts: How Forests Function*. Rutgers University Press, New Brunswick, NJ (2008).
3. Read. The ties that bind; Read. Plants on the web; van der Heijden et al. Mycorrhizal fungal diversity; Marsh et al. The role of *Equisetum*; Maser et al. *Trees, Truffles, and Beasts*.
4. This paragraph is based on Michael P. Amaranthus, Debbie Page-Dumroese, Al Harvey, and others. *Soil Compaction and Organic Matter Affect Conifer Seedling Nonmycorrhizal and Mycorrhizal Root Tip Abundance and Diversity*. U.S. Department of Agriculture, Forest Service, Pacific Northwest Research Station, Portland, OR, 1966. Research Paper PNW-RP-494; Daniel L. Luoma. *Monitoring of Fungal Diversity at the Siskiyou Integrated Research Site, with Special Reference to the Survey and Manage Species* Arcangeliella camphorata *(Singer & Smith) Pegler & Young*. Unpublished Final Report, Order #43-0M00-0-9008. Department of Forest Science, Oregon State University, Corvallis, OR, 2001; J. E. Smith, R. Molina, M. M. P. Huso, and others. Species richness, abundance, and composition of hypogeous and epigeous ectomycorrhizal fungal sporocarps in young, rotation-age, and old-growth stands of Douglas-fir (*Pseudotsuga menziesii*) in the Cascade Range of Oregon, U.S.A. *Canadian Journal of Botany* 80, (2002):186–204.
5. Peter Högberg, Anders Nordgren, Nina Buchnamm, and others. Large-scale forest girdling shows that current photosynthesis drives soil respiration. *Nature*, 411 (2001):789–792.

6. Eric D. Forsman, E. Charles Meslow, and Howard M. Wight. Distribution and biology of the spotted owl in Oregon. *Wildlife Monographs,* 87 (1984):1–64; Jack Ward Thomas, Eric D. Forsman, Joseph B. Lint, and others. *A Conservation Strategy for the Northern Spotted Owl: Report of the Interagency Scientific Committee to Address the Conservation of the Northern Spotted Owl.* U.S. Government Printing Office, Washington, DC (1990); Andrew B. Carey, Janice A. Reid, and Scott P. Horton. Spotted owl home range and habitat use in southern Oregon coast ranges. *Journal of Wildlife Management,* 54 (1990):11–17; Chris Maser. *Mammals of the Pacific Northwest: From the Coast to the High Cascade Mountains.* Oregon State University Press, Corvallis, OR (1998).
7. Zane Maser, Chris Maser, and James M. Trappe. Food habits of the northern flying squirrel (*Glaucomys sabrinus*) in Oregon. *Canadian Journal of Zoology,* 63 (1985):1084–1088; Chris Maser, Zane Maser, Joseph W. Witt, and Gary Hunt. The northern flying squirrel: A mycophagist in southwestern Oregon. *Canadian Journal of Zoology,* 64 (1986):2086–2089.
8. C. Y. Li, Chris Maser, and Harlan Fay. Initial survey of acetylene reduction and selected mircoorganisms in the feces of 19 species of mammals. *Great Basin Naturalist,* 46 (1986):646–650; C. Y. Li, Chris Maser, Zane Maser, and Burce Caldwell. Role of three rodents in forest nitrogen fixation in western Oregon: Another aspect of mammal-mycorrhizal fungus-tree mutualism. *Great Basin Naturalist,* 46 (1986):411–414.
9. Chris Maser, James M. Trappe, and Ronald A. Nussbaum. Fungal-small mammal interrelationships with emphasis on Oregon coniferous forests. *Ecology,* 59 (1978):779–809; Chris Maser, James M. Trappe, and Douglas Ure. Implications of small mammal mycophagy to the management of western coniferous forests. *Transactions of the 43rd North American Wildlife and Natural Resources Conference,* 43 (1978):78–88; Daniel L. Luoma, James M. Trappe, Andrew W. Claridge, Katherine M. Jacobs, and Efren Cázares. Relationships among fungi and small mammals in forested ecosystems. In: *Mammalian Community Dynamics in Coniferous Forests of Western North America: Management and Conservation,* Cynthia J. Zable and Robert G. Anthony (eds.), Chap. 10. Cambridge University Press, New York, 2002.
10. G. P. Calvo, Zane Maser, and Chris Maser. A note on fungi in small mammals from the *Nothofagus* forest in Argentina. *Great Basin Naturalist,* 49 (1989):618–620.
11. Andrew W. Claridge, M. T. Tranton, and R. B. Cunningham. Hypogeal fungi in the diet of the long-nosed potoroo (*Potorous tridactylus*) in mixed-species and regrowth eucalypt stands in south-eastern Australia. *Wildlife Research,* 20 (1993):321–337; Maser et al. *Trees, Truffles, and Beasts.*
12. Maser. *Mammals of the Pacific Northwest.*
13. David Schimel, Jerry Melillo, Hangin Tain, A. David McGuire, and others. Contribution of increasing CO_2 and climate to carbon storage by ecosystems in the United States. *Science,* 287 (2000):2004–2006.
14. Owen B. Toon. How pollution suppresses rain. *Science,* 287 (2000):1763–1764.
15. Daniel Rosenfeld. Suppression of rain and snow by urban and industrial air pollution. *Science* 287 (2000):1793–1796.
16. Josef Hebert. Mining, electric power account for most toxics. *Corvallis Gazette-Times* (May 12, 2000).
17. Rosenfeld. Suppression of rain and snow.

18. R. J. Esher, D. H. Marx, S. J. Ursic, R. L. Baker, L. R. Brown, and D. C. Coleman. Simulated acid rain effects on fine roots, ectomycorrhizae, microorganisms, and invertebrates in pine forests of the southern United States. *Water, Air, and Soil Pollution*, 61 (1992):269–278.
19. The discussion of nitrogen is based on Charles E. Little. Report from Lucy's Woods. *American Forests* (March/April 1992):25–27,68–69.
20. The discussion of streams and acidification is based on Geoff Wilson. A flush of acid. *Northern Woodlands*, 11(2004):17.
21. The discussion of red spruce and acidity is based on Paul Schasberg. Red spruce feels the cold. *Northern Woodlands*, 11(2004):44–45.
22. The foregoing discussion of mycorrhizal fungi is based on Li et al. Role of three rodents; Chris Maser. *Sustainable Forestry: Philosophy, Science, and Economics*. St. Lucie Press, Delray Beach, FL (1994); Maser. *Mammals of the Pacific Northwest*; Chris Maser. *Our Forest Legacy: Today's Decisions, Tomorrow's Consequences*. Maisonneuve Press, Washington, DC (2005).
23. Discovery Channel. Machines and Engineering. http://www.discoverychannel.co.uk/machines_and_engineering/tunnels/index.shtml (accessed December 15, 2008).
24. Michael Grunwald. An agency of unchecked clout. *The Washington Post* (September 10, 2000).
25. The discussion of the U.S. Army Corps of Engineers is based on Jim Robbins. Engineers plan to send a river flowing back to nature. *The New York Times* (May 12, 1998); Michael Grunwald. The corps' divided mission. *The Oregonian* (February 17, 2000); Michael Grunwald. More powerful than a river. *The Washington Post*. In: *Albany (OR) Democrat-Herald, Corvallis (OR) Gazette-Times* (November 23, 2000); Frederic J. Frommer. Groups identify "wasteful" corps water projects. *Corvallis Gazette-Times* (March 3, 2000); Amalie Young. Court: Snake dam operation violates Clean Water Act. *Corvallis Gazette-Times* (February 17, 2001).
26. John Howard. Crews struggle to save California levees. *Corvallis Gazette-Times* (January 7, 1997).
27. California crews bolster levees; drinking water faces peril. *The New York Times* (January 7, 1997).
28. Levy failure causes new floods in California. *Corvallis Gazette-Times* (January 11, 1997).
29. John Howard. Receding floodwaters reveal fields of dead cows, horses. *Corvallis Gazette-Times* (January 10, 1997).
30. Matthew Yi. Brimming reservoirs keep California flood threats alive. *Corvallis Gazette-Times* (January 10, 1997).
31. Scott Faber. Get people off of nation's floodplains. *Corvallis Gazette-Times* (January 21, 1997).
32. Ibid.
33. Ibid.
34. Luna B. Leopold. Ethos, equity, and the water resource. *Environment*, 2 (1990):16–42.
35. Richard Manning. The oil we eat. *Harper's Magazine* (February 2004):37–45.
36. Ibid.
37. The preceding two paragraphs on bees and pollination are based on Janet N. Abramovitz. Learning to value nature's free services. *The Futurist*, 31 (1997):39–42.

38. The preceding two paragraphs on the loss of honeybees are based on Steve Newman. Earthweek: A diary of the planet. *Albany (OR) Democrat-Herald, Corvallis (OR) Gazette-Times* (June 6, 1999).
39. The foregoing general discussion of Nature's services is based on Stephen L. Buchmann and Gary Paul Nabhan. *The Forgotten Pollinators*. Island Press, Washington, DC (1997); Gretchen C. Daily, Susan Alexander, Paul R. Ehrlich, Larry Goulder, Jane Lubchenco, and others. Ecosystem services: Benefits supplied to human societies by natural ecosystems. *Issues in Ecology*, 2 (1997):1–16.
40. Barbara Nussbaum. Ubuntu. *Resurgence*, 221 (2003):13.
41. Sally Pomme Clayton. Thread of life. *Resurgence*, 221 (2003):29.
42. The preceding paragraphs regarding hunter-gatherer lifestyle are based on John Gowdy. Introduction. In: *Limited Wants, Unlimited Means*, John Gowdy (ed.), xv–xxix. Island Press, Washington, DC, 1998.
43. Leonard W. Moss. Observations on "The Day of the Dead" in Catania, Sicily. *The Journal of American Folklore*, 76 (1963):134–135.
44. W. J. Elliot, C. H. Luce, R. B. Foltz, and T. E. Koler. Hydrologic and sedimentation effects of open and closed roads. *Natural Resource News*, 6 (1996):7–8.
45. Richard T. T. Forman and L. Alexander. Roads and their major ecological effects. *Annual Review of Ecology and Systematics*, 29 (1998):207–231; Stephen C. Trombulak and Christopher A. Frissell. Review of ecological effects of roads on terrestrial and aquatic communities. *Conservation Biology*, 14 (2000):18–30; Richard T. T. Forman. Estimate of the area affected ecologically by the road system in the united states. *Conservation Biology*, 14 (2000):31–35.
46. Jonathan L. Gelbrad and Jayne Belnap. Roads as conduits for exotic plant invasions in a semiarid landscape. *Conservation Biology*, 17 (2003):420–432; W. Schmidt. Plant dispersal by motor cars. *Vegtatio*, 80 (1989):147–152; David K. Person, Matthew Kirchhoff, Victor Van Ballenberghe, and others. *The Alexander Archipelago Wolf: A Conservation Assessment*. Pacific Northwest Research Station, Portland, OR (1996). USDA Forest Service General Technical Report. PNW-GTR-384.
47. Logging-road problem bigger than thought. *Corvallis Gazette-Times* (January 22, 1998).
48. Bill Bishop. To reduce congestion, don't build more roads—close 'em. *Corvallis Gazette-Times* (May 20, 1998).
49. The discussion pertaining to Bishop's comments are based on Bishop. To reduce congestion.
50. The preceding two paragraphs discussing road closures are based on Peter Headicar. Traffic in towns. *Resurgence*, 197 (1999):22–23; John Whitelegg. Sorry lorries. *Resurgence*, 197 (1999):28–29.
51. James O. Wilson and James Howard Kunstler. The war on cars. *Slate* (1998). http://www.slate.com/id/3670/entry/24044/ (accessed January 22, 2009).
52. Bill Bishop. To reduce congestion.
53. Peter Headicar. Traffic in towns.
54. The preceding two paragraphs on road building are based on Jay Walljasper. Asphalt rebellion. *Resurgence*, 193 (1999):11; Jane Silberstein and Chris Maser. *Land-Use Planning for Sustainable Development*. Lewis Publishers, Boca Raton, FL (2000).

Section III

Shared Relationships Between Everyforest and Everycity

The line between the end and the beginning is sometimes hard to discern, like the line separating the sand from the sea. They seem to run together for a while, and what we think is an ending often becomes a new beginning.

Musician Bill Gaither*

* Bill Gaither and Ken Abraham. *It's More Than the Music: Life Lessons on Friends, Faith, and What Matters Most.* Published by Warner Faith, Nashville, TN. (2003) 320 pp.

8

Cumulative Effects, Lag Periods, and Thresholds

> I would rather live in a world where my life is surrounded by mystery than live in a world so small that my mind could comprehend it.
>
> **American Clergyman Henry Emerson Fosdick[1]**

Community is rooted in a sense of place through which the people are in a reciprocal relationship with their landscape. As such, a community is not simply a static place within a static landscape but rather is a lively, ever-changing, interactive, interdependent system of relationships. Because a community is a self-organizing system, it not only incorporates information but also changes its environment. Thus, as the community in living alters the landscape, so the landscape in response alters the community.

Reciprocity is the self-reinforcing feedback loop that either extends or withholds sustainability from a community and its landscape. We therefore create trouble for ourselves when we confuse order with control. Although a relative sense of freedom and order is a partner in generating a viable, autonomous community, it is, nevertheless, an open system wherein continual change occurs in the "invisible present."

The invisible present is our inability to stand at a given point in time and see the small, seemingly innocuous effects of our actions as they accumulate over weeks, months, and years. Although we can all detect change, it does not always register in our consciousness until it has reached a certain magnitude. It is an unusual person who can sense, with any degree of precision, the changes that occur over the decades of their life. At this scale of time, we tend to think of the world in some sort of "steady state," and we typically underestimate the degree to which change has occurred as well as the imprecision with which we remember its details. We are unable to directly sense slow alterations, and we are even more limited in our abilities to interpret their relationships of cause and effect. The subtle processes that act quietly and unobtrusively over decades reside cloaked in the invisible present.[2]

The invisible present is the scale of time that encompasses our lives when our responsibilities to the children, present *and* future, are most evident. And, hidden within this scale of time are the interwoven threads of cause and effect that send unwise decisions reverberating throughout city and forest, as myriad, historical lessons teach—lessons we continually ignore at our collective peril, lessons that include the triad of cumulative effects, lag periods, and thresholds.

Cumulative effects, gathering themselves in the shadow of our consciousness, suddenly become visible. But, then it is too late to retract our decisions and actions, even if the outcome they cause is decidedly negative with respect to our intentions. I say "in the shadow of our consciousness" because of the aforementioned pervasive problem of our inability to visually perceive the gradual changes taking place in the invisible present. So it is that the cumulative effects of our activities compound unnoticed ("lag period") until something in the environment shifts dramatically enough for us to become consciously aware of it. That shift is defined by the "threshold," beyond which the system as we know it, suddenly, visibly becomes something else.

This said, the invisible present, including the lag period, is as much a matter of awareness as anything. They are both *invisible* as long as we ignore them, but should we wish, we could measure many of the minute changes that constitute the cumulative effects and thereby progressively nullify the lag period by making the "unseen" at least partially visible, be it positive or negative.

The Triad in Everyforest

Although fire is a physical process through which Nature originally designed forests in the western United States, that is not the way Gifford Pinchot (first chief of the U.S. Forest Service) saw it as he rode through parklike stands of ponderosa pine along the Mogollon Rim of central Arizona in the year 1900. It was a warm day in June as Pinchot rode his horse to the edge of a bluff overlooking the largest continuous ponderosa pine forest in North America. The pine-scented forest, without a logging road to scar the ground or a chain saw to tear the silence, was a sight to behold.

"We looked down and across the forest to the plain," Pinchot wrote years later. "And as we looked there rose a line of smokes. An Apache was getting ready to hunt deer. And he was setting the woods on fire because a hunter has a better chance under cover of smoke. It was primeval but not according to the rules."[3]

The forest over which Pinchot gazed on that June day in 1900 was three to four hundred or more years old, trees that had germinated and grown throughout their lives in a regime characterized by low-intensity surface fires sweeping repeatedly through their understory. These fires had occurred every few years or so, burning dead branches, stems, and needles on the ground and simultaneously thinned clumps of seedlings growing in openings left by vanquished trees. Although fire had been a major architect of the parklike forest of stately pines that Pinchot admired (a positive, cumulative effect), he did not understand the significance of fire in designing the forest or the indigenous peoples' role in perpetuating them.

What is interesting is that *Sunset Magazine* contained an article in 1910 that recommended to the fledgling Forest Service that it use the indigenous American's method of setting "cool fires" in the spring to keep the forests open, consume accumulated fuel, and in so doing protect the forest from catastrophic fire.[4] Unfortunately, that recommendation came the same year that, in the space of "two days in Hell," fires raced across three million acres in Idaho and Montana and killed eighty-five firefighters in what is called the "Big Blowup." It would be ten years after the Big Blowup before fires in western forests and grasslands were effectively controlled.[5] By 1926, the objective was to control all fires before they grew to ten acres (four hectares) in size. And, a decade later, the policy was to stop all fires by 10 a.m. on the second day.[6] For decades thereafter, the U.S. Forest Service was dedicated to putting out all fires. Such a policy is misplaced, however, because it ignores the primary cause of forest fires.

The reaction of the Forest Service is not surprising when one considers that most people prefer the devil they know to the devil they do not, which is but saying that the "terrible known" (a catastrophic fire) is often more comfortable than the unknown (setting "cool fires" like the indigenous Americans), even if the unknown promises to be better. People thus chart a course by consciously avoiding charting a course, meaning that a manageable situation is neglected until it is thoroughly out of hand.

In contraposition, and given enough time without human intervention, as stated earlier, virtually all forest ecosystems evolve toward a critical state in which a minor event sooner or later leads to a major event, one that alters the ecosystem in some fundamental way. As a young forest grows old, it converts energy from the sun into living tissue that ultimately dies and accumulates as organic debris on the forest floor. There, through decomposition, the organic debris releases the energy stored in its dead tissue. In this sense, a forest equates to a dissipative system in that energy acquired from the sun is dissipated gradually through decomposition or rapidly through fire. Before fire suppression was instigated, fires had burned frequently enough to generally control the amount of energy stored in accumulating deadwood by burning it up and so protected a forest for decades, even centuries, from a catastrophic, killing fire.

Although Pinchot knew about fire, he was convinced it had no place in a "managed" forest, so fire was to be vigorously extinguished, especially after the 1910 Big Blowup. In addition, conventional wisdom dictated that ground fires kept forests "understocked," and more trees could be grown and harvested without fire. Further, surviving trees were often scarred by the fires, like the ones Pinchot had seen in Arizona, allowing decay-causing fungi to enter the stem, reducing the quantity and quality of harvestable wood. Finally, any wood not used for direct human benefit was thought an economic "waste." Here, we need to keep Pinchot's two ideas in mind: Fire has no place in a managed forest, and what is not used to the material benefit of society is an economic waste.

In Pinchot's time and place in history, *he was correct and on the cutting edge*, especially since the ecological problems caused by such thinking were unbeknownst at that time. Nevertheless, incorporation of these ideas into forestry began to take their toll. Only now, after decades of suppressing fires and planting blocks of even-aged trees across whole landscapes (decades of negative, cumulative effects compounding unobtrusively throughout the lag period) has the significance—the threshold—of the resulting, deleterious changes in the composition, structure, and function of forests become evident, such as fires increasing in numbers, destructive intensity, acres (hectares) burned, and severity in damage to human property.[7]

At this point, linear, commodity-oriented thinking entered into the profession of forestry in the United States, and Pinchot's utilitarian conviction that fire had no place in a managed forest became both the mission and the metaphor of the young agency that he built. Managed, in this sense, has come to mean any forested acre where someone perceives an economic interest in the trees. It was this notion of linearity—of economic waste if a potential commodity was not used by humans for the demonstrable benefit of humans—that introduced the "invisible present" into the profession of forestry so long ago. How might cumulative effects, lag periods, and thresholds operate in the invisible present within a city?

The Triad in Everycity

The triad of cumulative effects, lag periods, and thresholds has been affecting human settlements ever since the inception of the first one. In those early days, on the other hand, a settlement was more closely linked to its immediate landscape and so was more readily, swiftly, and often dramatically affected by the alterations people made in their surroundings, as illustrated next.

Yesteryear's Town

The discovery of rich deposits of silver in 1870 marked the beginning of the mining industry in Silver City, New Mexico. The area around the city was mined extensively, and with a focus solely on mining, the forest was completely cut over between 1870 and 1887 to provide fuel for steam boilers at mines, to build structures for mining, and to feed household fires.

In addition, the communal grazing of livestock around the city was indiscriminate between 1870 and 1908. During the dry years prior to 1900, according to old-timers, as many as 1,500 head of cattle would graze in close proximity to the city, and that does not take into account the sheep, goats, mules, burros, horses, and even swine in some places. Consequently, the town's immediate surroundings became badly overgrazed, and since most of the available forest had already been cut, there was practically no groundcover to hold the soil in place.

The cumulative effects of logging and livestock grazing occurred within the concealing shadow of the invisible present. During this time, Main Street was the city's primary north-south thoroughfare, the principal artery of commerce, the point of arrivals and departures, and the social center of town. The center of the street also provided moderate drainage as it was two to three feet lower than either side. Then came the threshold, and the people of Silver City were left with a conscious and painful reminder of Nature's awesome power.

It began with torrential rains on July 21, 1895. With no vegetation left around the town to dissipate the rain's pounding energy so the water could be retained long enough to infiltrate the soil, it headed into Silver City. Pouring off the denuded land, the runoff created a monstrous gully out of Main Street, the bottom of which was 35 feet (10.6 meters) below the previous afternoon's street level. Subsequent floods climaxed in a two-day assault on the denuded water catchment in August 1903, scouring the gully down to bedrock, 55 feet (17 meters) below Main Street's original level of traffic, and the erosion continued some 15 miles (24 kilometers) south of town.

What happened in Silver City is a prime example of an irreversible result, in this case caused by indiscriminate mining, logging, and livestock grazing. Although the ecological damage to the city's landscape has been partially healed, it is not yet nearly as stable as it was prior to 1870. And as of 1980, a project was under way to establish a recreational walk along the sides of the gully known as the Big Ditch.[8]

What happened in Silver City seems pretty clear because the town was small and the evidence is documented, both in photographs and by the narratives of people who experienced it. Other cause-and-effect relationships, however, are neither as immediately dramatic nor as clearly obvious because they either take place as a planned event within a city and so are commonplace experiences or happen gradually over many decades, often exceeding a human life span before the threshold makes its visible appearance.

Today's City

Perhaps one of the simplest ways of examining cumulative effects, lag periods, and thresholds in today's city is to watch a planned, commonplace event—a tall building under construction. In this case, the initial threshold is the decision to construct the building, a choice that most often acts as an irreversible point in time. With that threshold crossed, the contractors gather at the site. A protective fence is erected, all of which is readily noticeable to people passing by it. Whether they take heed of these events is another matter.

Then, it begins. A huge machine commences to take great "bites" out of the soil and load them onto a truck that hauls them away—bite after bite after bite, day after day after day, truckload after truckload after truckload. Each bite adds its cumulative effect of removing soil that was begun with the first bite. Gradually, the hole deepens, and the effect becomes increasingly irreversible. At some point, a passerby, who has walked alongside the deepening hole for weeks, will suddenly stop and think, "Geez, what a hole!" For that person, the cumulative effects (of all those bites of soil loaded and hauled away) finally become a visible change because the person's threshold of awareness was crossed and the hole suddenly "became visible." Yet to others, it may still be invisible. A mechanically inclined child, on the other hand, may have noticed the first big piece of equipment and has come back as often as possible to watch the progress of the deepening hole.

Finally, after many weeks of digging and hauling soil, labor begins on the foundation, way down in the bottom of the pit. In this case, the operations are largely out of sight for most passersby and so are mostly unnoticed. Moreover, constructing the foundation for a fifty-story building is not only hidden from the street level in the beginning but also proceeds at a snail's pace from a visual point of view. Consequently, it is not very exciting and so elicits scant curiosity or attention.

As the days turn into weeks and then into months, the foundation creeps streetward and is occasionally noticed by a passerby. But, whenever the street level is reached, the threshold is crossed for many people because thereafter the floors seem to rise at the rate of one every couple of weeks. Once the building has been rough framed, on the other hand, it again passes into the realm of the invisible present because the snail's pace again takes hold.

Although the construction of a single building does not overtly alter an entire city as a functioning system, it does influence it. Here, one has only to remember the Twin Towers in New York City and all the people who were affected by their collapse to understand the impact a single structure can have on people who get to know it intimately. There is yet another aspect to a single building within a city, regardless of its stature. Namely, each building helps to create and simultaneously define the city as people come to know it. As well, each building has an effect on the city's environment and beyond. To put in perspective the cumulative effects today's towns and cities are having on our home planet as a whole, let us use a common activity, one most people perform unconsciously, with respect to the hidden ramifications of their behavior—namely, lawn care.

The Cumulative Effect of Today's Cities

Most city folks, especially suburbanites, spend much money to hire someone to fertilize their lawns; mow them for a neat appearance; and spray weed killer on them to keep a pure monoculture of grass—especially true of golf courses. But, all these chemicals ultimately go somewhere because they do not stay in people's yards or on golf courses. Where do they go? In Corvallis,

Oregon, as in many cities, they go down through the soil and eventually into a local creek.

Dixon Creek, flowing through a part of Corvallis, drains 4.8 square miles (12.4 square kilometers) or 3,041 acres (1,231 hectares), including upstream use that is 18 percent agriculture, 20 percent forest, and 62 percent residential. The creek, only a few miles (kilometers) long and almost entirely within the city limits, flows past churches, parks, homes, and even through the campus of a local high school before emptying into the Willamette River, which in turn empties into the Columbia River that empties into the Pacific Ocean. In its flowing through the invisible present of many decades, Dixon Creek has accumulated traces of nine pesticides and herbicides, as well as fecal coliform bacteria (*Escherichia coli*), according to a 1998 study of water quality in the Willamette River basin of western Oregon conducted by the U.S. Geological Survey.

The pesticides found in Dixon Creek are listed in descending order of concentration:

- *Dichlobenil,* a herbicide
- *Tebuthurion,* a herbicide not known to be hazardous to aquatic organisms
- *Diazinon,* an insecticide toxic to fish
- *Carbaryl,* an insecticide moderately toxic to aquatic organisms
- *Prometon,* a herbicide
- *Metolachlor,* a herbicide moderately toxic to both cold- and warm-water fish
- *Atrazine,* a herbicide slightly toxic to fish and other pond life
- *Desethylatrazine,* a herbicide slightly toxic to fish and other pond life
- *Simazine,* a herbicide of low toxicity to aquatic species

In addition, 20 percent of the water samples from Dixon Creek were hotter than 68° Fahrenheit (20° Celsius) during summer's low water flow, which is warm enough to cause increased stress on fish.[9] And, this is just a small example, one that can be generalized to streams and rivers throughout the United States and much of the world.

Such poisons as these affect the environment both through differences in their chemical makeup and in the way they kill; what they kill; in what concentrations they can kill; how they move through the food chain, killing as they go; how and to what they spread their deleterious but sublethal effects and the deadly synergism of their chemical interactions when they come in contact with one another. Ponder also how, through time, they alter the ecosystem in which their respective effects become manifest—although not necessarily understood, or even noticed, by humanity.

Beyond that, they collect unnoticed in the ditches, including street-side gutters, and polluted creeks in and around the myriad small towns and large cities, all of which add their array of poisons to the aquatic environment. And, this does not include the nonpoint source, chemical pollution that leaches into the groundwater from intensive agriculture (including feedlots for cattle, pigs, and poultry) and intensive forestry (tree farms).

Now, the arsenal of silent, "invisible" toxins is mighty indeed as it enters the ditches, streams, rivers, estuaries, and oceans of the world, where all pollution concentrates over time as a continual progression of cumulative effects because oceans have no outlets. Oceans lose their water through evaporation that concentrates the remaining chemicals even more because there is no way to dilute them without pure water flowing into the oceans and outlets through which to flush them.

The oceans' inability to flush themselves as do ditches, streams, and rivers causes me to wonder: How many of the chemicals introduced into our environment are toxic and to what degree? How many of the chemicals are in fact biodegradable, breaking down into harmless components over what period of time? How many of the chemicals recombine and result in "new" toxic compounds that may or may not be biodegradable? I have yet to receive a single satisfactory answer when these questions are posed, even to chemists. And when, I wonder, will the threshold be crossed and the oceans begin to die?

Protect the soil and we protect the humble ditch; protect the ditch and we protect the stream, river, estuary, ocean, and the cities that rely on them. Defile the soil and we defile the ditch; defile the ditch and we defile the stream, river, estuary, ocean, and the cities that rely on them. Which of these alternatives we choose is up to us, the adults, in the invisible present. This being so, we can always choose to honor the humble and thereby protect the mighty for it is ordained in the nature of things that water always seeks the lowest level and thereby knows where it is going—back to the sea. Whether it takes days, months, or years makes no difference.[10] But then, the invisible present being what it is, most people, especially product-oriented thinkers, do not even contemplate ditches and their connections to the cities and ultimately to the oceans of the world. In fairness, part of the reason cumulative effects tend to be ignored is the invisible present of the lag period in both forests and cities.

Summation

Although Chapter 8 was an examination of cumulative effects that take place in the invisible present, and are everywhere an ongoing component of Nature's biophysical systems, I was concerned only with those of human origin. Cumulative effects can be either positive or negative, depending on how one interprets the outcome. Moreover, cumulative effects compound out of sight over time (the lag period) until, having sufficiently altered a system, they cross a threshold and become visible, at which time they are totally beyond even the slightest degree of reversibility.

The invisible present does have a positive note, however, and that is aging as an archive of history, both in forests and cities, which is the subject of Chapter 9.

Notes

1. Henry Emerson Fosdick. Quotations and Passages of Faith. http://www.livinglifefully.com/faith2.html (accessed January 22, 2009).
2. John J. Magnuson. Long-term ecological research and the invisible present. *BioScience*, 40 (1990):495–501.
3. Gifford Pinchot. *Breaking New Ground*. Harcourt, Brace and Co. Inc., New York (1947).
4. George L. Hoxie. How fire helps forestry. *Sunset*, 34 (1910):145–151.
5. Stephen J. Pyne. *Year of the Fires: The Story of the Great Fires of 1910*. Viking Press, New York (2001); Michael Williams. *Americans and Their Forests: A Historical Geography*. Cambridge University Press, New York (1989).
6. Tom Kenworthy. Prevention efforts still missing mark after 2 years and $6 billion. *USA Today* (August 22, 2002).
7. Frederick J. Deneke. Forestry: An evolution of consciousness. *Journal of Forestry* 96(1998):56; Wally W. Covington and M. M. Moore. Post-settlement changes in natural fire regimes and forest structure: Ecological restoration of old-growth ponderosa pine forests. *Journal of Sustainable Forestry*, 2 (1994):153–182; Wally W. Covington and M. M. Moore. Southwestern ponderosa pine forest structure: Changes since Euro-American settlement. *Journal of Forestry*, 92 (1994):39–47; Thomas W. Swetnam. Forest fire primeval. *Natural Science*, 3 (1988):236–241; Thomas W. Swetnam. Fire history and climate in the southwestern United States. In: *Effects of Fire in Management of Southwestern Natural Resources*, J. S. Krammers (tech. coord.), 6–17. Rocky Mt. Research Station, Fort Collins, CO, 1990. USDA Forest Service General Technical Report RM-191; Stephen J. Pyne. Where have all the fires gone? *Fire Management Today*, 60 (2000):4–6.
8. The foregoing discussion of Silver City is based on J. T. Columbus. Watershed abuse—the effect on a town. *Rangelands*, 2 (1980):148–150.
9. The foregoing discussion of pollution in the waters of Dixon Creek is based on: Scott MacWilliams. What's in the water? *Corvallis Gazette-Times* (February 14, 1998).
10. Chris Maser. The humble ditch. *Resurgence* 172 (1995):38–40.

9

Age as an Archive of History

> Endless invention, endless experiment,
> Brings knowledge of motion, but not of stillness …
> Where is the Life we have lost in living?
> Were is the wisdom we have lost in knowledge?
>
> **T. S. Eliot**[1]

History is a reflection of how we see ourselves and thus goes to the very root of how we give value to things. Our vision of the past is shaped by, and in turn shapes, our understanding of the present—those complex and comprehensive images we carry in our heads whereby we decide what is true or false,[2] which is probably the best we can do if we see life the same as video artist Bill Viola: "Life is just a brief emergence from the great sea of time."[3]

Alas, "the lack of a sense of history is the damnation of the modern world," wrote author Robert Penn Warren,[4] a thought akin to that of German philosopher George Wilhelm Friedrich Hegel: "Experience and history teach … that people and governments have never learned anything from history."[5] Our apparent disregard for history is indeed tragic, as military historian Robert Epstein pointed out: "Ancient history has everything. There is nothing that can ever happen that won't have an echo from the classical past,"[6] an echo we in the present can hear *if we will only listen.*

My mother, who was born and raised in Cologne, Germany, grew up surrounded by old buildings and marble statues in the likeness of and dedicated to people of antiquity. Even some of the cemeteries I visited in Europe had pictures of the deceased on their headstones, often prints on glass or tintypes that depicted the era in which the people had lived. These were cemeteries with rare character and a sense of historical dignity as well as humanity.

Having grown up surrounded by these images of antiquity, my mother had a consummate love affair with history all her adult life, and some of the history that fascinated her the most was that of Egypt. Therefore, while working in Egypt in 1963–1964, I went into the tiny inner sanctum of Abu Simbel Temple before it was moved. There, I filled a little glass vial with sand. It was the best possible gift I could give my mother because it fueled her vivid imagination for years.

The sand had survived the millennia as it scurried over the desert compelled by ever-restless winds. At length, some grains came to rest within

Abu Simbel, where they might have been touched around 2,050 years ago by Rameses II (1292–1225 BCE) as he entered the tiny room. What is important is that a small vial of sand transformed Abu Simbel Temple, on the other side of the world, from a mere historical abstraction into a personal dimension for my mother.

I, on the other hand, was interested in the graffiti, the names and dates of British soldiers carved into the interior walls of the temple in 1832, during Kitchener's war with the Sudanese—names that were a mere 131 years old in 1963. What the sand from Abu Simbel Temple did for my mother, the graffiti did for me when I touched it. In both cases, a personal identity was formed with a historical site that would not have happened otherwise—an identity my mother has already carried to her grave, and I will carry to mine.

Yet, this kind of personal identity can also be a liability, as pointed out by William Blake, who observed that, if you want to pull down a civilization, you should first attack its arts.[7] By this, Blake meant the vehicles of its inspiration and ideals: museums, libraries, schools, and other cultural symbols, something as complex as a bridge or as simple as a tree. (If you want to see what some of these sites are, look at Colin Wilson's marvelous book, *The Atlas of Holy Places and Sacred Sites*."[8])

Scattered throughout various parts of the world there still exists an old tree in the middle of the square around which village life revolves. It is a quaint meeting place in which neighbors form bonds with one another, children play games, women visit about the affairs of life, and men discuss work and politics. It is a place where old and young mingle in a way that bridges the generations in the eternal flow and ebb of village life. And, it is a place where children still experience an unstructured and noncompetitive setting in which to play and imagine, with the safety of their parents close at hand. As such, a village commons is far more than simply a public space around a tree. It is the center, where the life of true community blossoms because it has the scale of a human face, a human touch, and the continuity of time. Destroy the tree (the symbolic heart of community), and the village story begins to wither and fade as though suffocated by a scorching desert wind.

And that is exactly what happened during the 1990s when the bridges in Kosovo were blown up. The people who built those bridges more than a century ago had personal stories enveloped in hopes and fears, in love and pain, to say nothing of the interpersonal harmony or strife with which the bridges were built. As well, the people who crossed the bridges through all the decades each had their own story. Collectively, the people conferred onto the bridge a personality through their daily contact with it; in return, the bridge became the archive of those myriad stories that were such an integral part of Kosovo's historic soul, an irretrievable loss to a culture shattered by war.

Age as an Archive

Age is an archive of history wherein ancient trees and forests have much in common with historic buildings, districts, and cities. Both have humble beginnings (a seed; an idea, a small building), just like a tapestry, for which the weaver (whether Nature or a person) begins at the bottom with a single strand of fiber, the foundation of the image (ultimately, the design), and builds upward with the addition of each fiber. In this way, every living tapestry (whether a forest or a city) sees the birth of its foundation become progressively older as the ever-newer components are added, until that time when the beginning passes into antiquity.

When you plant a melon seed, it germinates, puts down roots, sends up shoots, and grows into a vine that adds flowers pollinated by bees. A flower, having served its purpose, withers, but its birthplace commences to produce a fruit. The fruit in turn grows and in so doing goes from unripe to ripe to overripe. Our task, yours and mine, is to pick the fruit at exactly the time of its perfection. Wait too long, and the melon returns to the soil from whence it came. This is one cycle of aging.

Spring is the time when the plants of a garden are young and soft and tender, their new leaves a bright green, burgundy, or gentle silver. By summer, the plants are mature, and their leaves are often coarse and dull, with simple holes and creative lacework eaten into them by myriad insects, pill bugs, and ever-hungry slugs. With the arrival of autumn, the leaves begin to wither in hot, drying winds. But, it is the cold north winds of approaching winter that cause them to break loose their bonds and joust and bounce their way to earth, where they disappear into the unknown from whence they came, into the atomic interchange of the soil they will enrich with their passing. This also is a cycle of aging.

Watching our body age is a profound, concrete example of cumulative effects, a lag period, and a threshold. Most of us were unaware of the long-term effects unwise, youthful actions would visit on our bodies as we aged. We felt immune to the effects of sun and wind, gathering day by day over months and years in the invisible present until the time arrived when the cumulative effects emerged from the unknown lag period of intervening years and showed themselves on the quality of our aging skin. Only then might we understand that the wisdom of old age cannot "cure" the ignorance of youth; that every abuse of our bodies is added to a tally sheet, the balance of which we must all pay—at times with interest. But, how does a building show age?

Like an old tree that has craggy bark and broken limbs, a wooden building in the country begins to turn silvery and to crack with sun and wind if boards are unpainted. With cracks come broken pieces through which the storm winds of winter can blow their chilling breath. Through the long count of years, it might sag a little because of loosened joints that allow it to creak in the wind, giving it a sense of life and foreboding to the imagination. The aging of a building in a city, however, is somewhat subtler, both within and without. What kind of changes might take place as a city building marks time?

First, let us look inside:

- The electrical wiring ages with the building, both in real time and because technology has resulted in a new and better kind of electrical wire that will be improved again sometime in the future.
- The plumbing system of old metal pipes has been outdated by plastic pipes that do not rust and eventually will be replaced by yet some other kind of pipe.
- The toxic lead-based paint that required turpentine to dissolve is replaced by a nontoxic, water-based latex paint that easily washes off in water.
- Wear caused by many years of human use becomes increasingly obvious, such as nicks in doorjambs, scratches on floors and walls, smudges about light switches, discoloring and oxidation of the paint, and so on.
- Solid wood construction has given way to plywood; then to chipboard, waferboard, and laminated beams, and so on.

Now let us look outside:

- Continual change takes place in the surrounding area, such as growth in the population, accompanied as it is by an increasing number of buildings with a more complex infrastructure as the town grows.
- The architectural styles of buildings transform through the years, from an artistic gift to the centuries to ever-more *modern* "crackerbox" designs with built-in obsolescence that translates into increasingly short life spans for a "throwaway" society.
- There are changes in human values with the growing modernization of a society from a life filled with solid objects of lasting beauty and grace to a life based on sound bites, bottom lines, discardable objects, and an ever-present demand for increased speed; these changes would seem to be a paradox in that they appear to reverse the aging process of social evolution, but they do not do so in fact

So it is that the cycle of aging advances the notion that to exist is to change; to change is to mature; to mature is to be endlessly entrained by time, the conveyor of age. The phenomenon of aging is equally true for a fine cheese, the excellence of a vintage wine, the historical record of a building, or the veneration of a tree. There is a caveat with trees, however.

Many people think that a large tree is necessarily an old tree, but that is not always true. There are circumstances under which a tree can grow quickly, reaching great size, and still be young in years. A truly "old" tree is old in years, hence physiologically old, not inevitably large in size. For example, I once took a group of people on a nature hike in Mount Rainier National Park in the state of Washington. As we climbed in elevation through large, stately Douglas-fir trees, someone said, "Wow, what a beautiful old-growth forest this is!" I said nothing until we had climbed high enough onto a rocky ridge to be standing in a stunted grove of true fir. As we looked over the tops of the small trees, I said, "Folks, welcome to the true ancient forest. These little trees are 600 years old, whereas the large Douglas-firs through which we walked earlier are a mere 200 to 300 years old. Old age in a forest is based on physiology, not chronology."

Nevertheless, as great size gives the illusion of old age in trees, the illusion of age is simulated with jewelry by adding fake patina in order to create an aura of mystery, the echoes from a distant past and, perhaps, a distant land somewhere beyond the horizon of imagination. So it is that our social-environmental tapestry is forever aging even as it is being created and re-created, repaired, and replaced. In this sense, an old tree is analogous to an old building in a city, and an old forest is analogous to a historic district in a city.

Age as an Archive in Everyforest

Now, let us visit briefly with the giant sequoia trees that inhabit scattered areas of central California on the west side of the Sierra Nevada mountain range. Between 5,000 and 4,000 years ago, some of the trees germinated and grew until they fell with age or were cut down early in the last century. Although giant sequoias still live, they are not so old; one tree, the General Sherman Tree in Sequoia National Park, was estimated to be 3,800 years old in 1968. It would have germinated 1,832 BCE and would have been 132 years old when, tradition says, a people called the Hyksos came from the east and conquered the Nile Delta at the end of the thirteenth dynasty in Egypt. The tree would have been 632 years old when the Trojan War was fought in 1,200 BCE and 1,056 years old when the first Olympic games were held in 776 BCE. When the Great Wall of China was built in 215 BCE, the tree would have been 1,617 years old. Thus, the tree would already have been 1,832 years old when Jesus was born in Bethlehem.

There is another tree, named bristlecone pine, that occurs today only on high peaks from Colorado to southern Utah, central and southern Nevada, southeastern California, and northern Arizona. There was a bristlecone pine in Great Basin National Park, Nevada, that was over 5,000 years old when it was cut down about 1990. If the bristlecone had been exactly 5,000 years old when it was cut, it would have germinated in 3,010 BCE and would have been 1,010 years old when the pyramids were built in Giza, Egypt, and the Semites conquered the third dynasty of Ur at the northern end of the Persian Gulf, forever ending Sumerian rule. The tree would have been 4,785 years old when the American Revolution began and 4,797 years old when the Constitution of the United States was written.

While we cannot speak with them per se, trees have a language through which they converse with us—their growth rings. Each year, a tree develops two growth rings, a light, relatively wide one in the spring and a darker, narrower one in summer. By counting the pairs of rings, a tree tells us its age.

Trees also tell us about changes in weather patterns through the analysis of their growth rings. When years are wet and the temperature mild, their rings are relatively wide, indicating good years. But, when summers and winters are dry, their narrow rings tell us about drought. In addition, charcoal in a growth ring imparts information about fire. Certain irregularities indicate that a bear stripped the bark with its claws in search of the nutritious cambium (the tree's living tissue), then scrapped it off with its lower incisor teeth.[9] Holes and dead areas of the rings speak of injury and insects.

This is only part of the story a tree can tell. So, it is fortunate that a few ancient trees are protected in national parks and other places. But, sacred trees secreted in the memories of people are not protected because memories are seldom deemed of sufficient merit in our competitive, economically driven culture to warrant honoring and protecting a potential commodity for its historic value, despite the fact that doing so would be a wise long-term spiritual decision for future generations.

The foregoing statement is especially true for those old forests around the world that the timber industry views as windfall profit because they have no investment in them other than their conversion into products and money. There are, nevertheless, many valid reasons for saving the old in both forest and city.

Age as an Archive in Everycity

A community's history must be delivered from one generation to the next if the community is to know itself throughout the passage of time. One landmark of that past resides in historic buildings. Moreover, an area can be designated a historic district when it has enough old buildings. In both cases, the buildings record a snapshot in a particular era of the community's history, each woven into the march of time, and so imparts a unique ambiance as a sense of the past. As well, artifacts from those who went before us are housed within a museum maintained by the historical society, the caretaker of shadows from bygone days.

In this sense, it is no small irony, wrote James Howard Kunstler, that during the greatest era of prosperity in the United States, the decades following World War II, only the cheapest possible buildings were constructed, including civic buildings.[10] To understand what he meant, compare any richly designed post office or city hall built at the turn of the twentieth century with today's dreary, unimaginative, concrete box counterparts.

Kunstler's point is a good one. When the United States was a far less wealthy nation (by monetary standards), things were built to endure because it would have seemed immoral, if not insane, in our great-grandparents' day to throw away hard-earned money and honest labor, as well as waste valuable resources future generations would need, on something guaranteed to disintegrate within thirty years.

The buildings erected in those earlier days paid homage to history through their design, including elegant solutions to the age-old problems posed by the cycles of weather and light. They paid respect to the future because they were consciously built to endure beyond the lifetimes of the people who designed and constructed them. Kunstler accounted for this continuum of past, present, and future in what he called "chronological connectivity."

Chronological connectivity, said Kunstler, is a fundamental pattern of the universe: "an understanding that time is a defining dimension of existence—particularly the existence of living things, such as human beings, who miraculously pass into life and then inevitably pass out of it."[11] It puts us in touch with the ages and connects us with a sense of eternity; it connects us with our place in humanity's story and so indicates that we are somehow part of an organism that is infinitely larger than ourselves.

The notion of chronological connectivity suggests that the large organism we help to create even cares about us, and that requires us, in turn, to respect ourselves and all life that will follow us, just as those who preceded us respected those who followed them. This notion is important, asserted Kunstler, who practiced no formal religion, because it puts us in touch with the holy, that which is at once humbling and exhilarating. Connectivity with the centuries of the past and the horizons of the future lead us in the direction of enchantment, grace, and sanity.[12]

But, if the continuity of a community's history is disrupted, the community loses its place in its own story and suffers an extinction of identity. Then, having lost its story, the community is destroyed from within because trust is withdrawn in the face of growing economic competition from increasingly transient members or outside "predators."

We have rejected both the past and the future since 1945, said Kunstler, a repudiation that is plainly manifest in our graceless buildings, each constructed to disintegrate within a few decades. This consciously built-in decline is euphemistically termed *design life,* which may last fifty years. Since today's buildings are expected to serve only our era, we seem unwilling to expend money or effort for their beauty, maintenance, or service as storytellers to the generations of the future.[13]

Nor do we care about those elegant solutions to the problems created by the cycles of weather and light; after all, we have such technology as electricity, central heating, and air conditioning. Thus, many new office buildings have windows that cannot be opened or virtually no windows at all, further isolating people from the flow and ebb of Nature, such as the weather. This process of disconnecting from the time continuum of the past, through the present, into the future and from the cycles of weather and light diminishes us spiritually, impoverishes us socially, and destroys the time-honored cultural patterns we call community. As an army major at Fort Leavenworth, Kansas, put it in reference to the electronic media, they are about "*now, now, now*, with all the depth of a credit card."[14] Unfortunately, the same can be said of most so-called developers.

No community today is untouched by the interplay between its traditional self (the richness of its original story based on quality, social well-being, and effectiveness) and the greater, more expedient industrial-commercial society (the "Wall Street" story focused solely on quantity, efficiency, and monetary bottom lines). It is hardly surprising, therefore, that conflicts over the value of place arise with increasing frequency between those members of a community who hold the traditional values of lifelong residents and those who hold the "modern" values of an increasingly transient population. In this sense, many communities are in transition between sets of values that must be carefully assessed in terms of human attitudes and the way land is used. When last I was in Japan in the early 1990s, for example, I felt a palpable division between that segment of the population represented by the older Shinto Japanese and the younger, westward-leaning, materialistic Japanese.

There is today an exploding need to find a common language and conceptual framework of mutual understanding about the sense of place and its historical significance. I say this not only because of what we, in the present generation, have to lose but also because of what future generations have to lose if we, who will become the stuff of antiquity in that distant time, are not mindful of the legacy we pass forward. In addition, the sense of community is being lost as citizens become increasingly reticent to think in terms of maintenance. We Americans seem eager to build but then begrudge providing the dollars necessary to maintain our highways and schools, let alone our downtowns, which we effectively abandon in favor of ever-newer, isolated, "disposable" shopping malls and housing developments.

One of the reasons so many townspeople are reluctant to spend money on maintenance is their lack of long-term commitment to the town where they reside while earning a living because they plan to move somewhere else upon retirement to spend the rest of their lives. With such thinking, why would they be committed to spending their hard-earned money on maintaining a place they are eagerly planning to leave?

What we neglect, we lose, be it a house, a street, a downtown, or a community itself. Communities are neither meant to be disposable nor designed in terms of planned obsolescence. This could be partially remedied if each member of a community would agree to dedicate an annual percentage of his or her time to do something that would improve the quality of their community.

Tithing a portion of one's time is the beginning of recognizing the difference between real wealth and money. Conventional money knows no loyalty to a sense of place, a local community, or even a nation, and so it flows toward a global economy wherein traditional social bonds give way to a rootless quest for the highest monetary return. The real price we pay for money, the real cost, is the hold it has on our sense of what is possible—the prison it builds around our imagination. We forget, according to Bernard Lietaer of the Center for Sustainable Resources at the University of California at Berkeley, that money is an artifact that we both conjured and designed. The irony in our forgetfulness is that money is now "like an iron ring we've put through our noses," and it is now leading us around. "I think," said Lietaer, that "it's time to figure out where *we* want to go—in my opinion toward sustainability—and then design a money system to get us there."[15]

Lietaer went on to say that, while textbooks on economics claim that people and corporations are competing for resources and markets, they are in reality competing for money and in so doing are exploiting resources and using markets.[16] "A more fascinating aspect of money," remarked author Caroline Myss, "is the fact that it can weave itself into the human psyche as a substitute for the life-force."[17] This is an astute observation because we make our private thoughts into public declarations in the way we spend money, just as we do when we speak or write.

In sum, age in a city is the archive of history that survives in the creations of the human mind and heart, the artifacts that withstand the ravages of time and thereby safeguard the inspiration that gave them birth in the human psyche. Further, a culture's definition of prosperity and poverty will usually determine to what extent individuals in a culture work to live or live to work; to what extent prosperity is tied to material wealth; and conversely, to what extent poverty is tied to spiritual impoverishment.

Why Save the Old?

Humanity's search for meaning in life is paralleled by a search for its beginning—the first thought, the first human word, the first culture, the oldest building, the most ancient script. All of these things act to help us countervail our sense of frailty, vulnerability, and mortality as we face the uncertainties that daily beset us because they help us to visualize our place in the chronicles of humanity's journey through the never-ending mysteries of life.

Another way we face the unknowns of life is to leave a legacy by which we are remembered. To an architect, that is a building; to a planner, a city; to an author, a book; to a film director, a motion picture; to an artist, a painting or a photograph; to a citizen, perhaps a tree planted in memory of a loved one. Whatever it is, it is meant to carry some part of us, some concrete evidence of our existence, into the future and thereby tell our personal stories throughout the ages as time wends its way into the unknown and unknowable.

And, there is yet another reason to save the old in both forest and city; that reason is to teach each new generation about the basic principles on which their modernity is based, lessons (both positive and negative) that are fundamental to their social-environmental sustainability. Although language is the ultimate conveyor of knowledge and the principles envisioned by humanity, it is also the oldest artifact in the human psyche—the single, tangible link to the far memory of our species. But, language to most people is an abstraction that needs to be augmented by concrete examples of the idea, thing, image, or function expressed by words; therein lies what is perhaps the greatest value of saving the old—the ability to see, touch, smell, and at times, even taste whatever it is.

The Value of Saving Old Forests

Although there are many reasons for saving old forests, five are briefly discussed, all of which are related in some way to a counterpart in a city:

1. They are unique in time and space; once gone, they are irreplaceable.
2. They are a living library wherein we can study the biophysical principles of Nature's design elements that ultimately govern the functional dynamics of both forest and city.
3. Nature's biophysical principles function perfectly whether or not we understand them.
4. The living trees archive their knowledge of the past, which we can read, but only if the trees remain in Nature's living library, where they will become even more valuable for the education of future generations because they are continually recording environmental events.
5. An old forest is a metaphorical mirror of what it means to be separate individuals *and* simultaneously a universal people.

1. Old forests are both magnificent in stature and unique in a world of time and space; once gone, they are irreplaceable. Ancient trees inspire spiritual renewal in many people and are among the rapidly dwindling living monarchs of the world's forests. Old forests are the oldest living beings on Earth and thus form a tangible link with the past and provide a spiritual ground in the present. Saving them is to honor and protect the continuity of generations.

What would happen to our sense of continuity, our sense of spiritual grounding in the dynamic, unpredictable universe if all the remaining commercially available old forests were liquidated for short-term profits? How would you feel if these centuries-old living beings where suddenly replaced with trees that were not allowed to grow much older than a person? How would you feel if you knew that the oldest living beings on Earth had all been converted into money, and you would never again see them? By the same token, your children, your children's children, and their children's children would never have a chance to see them.

Old forests are unique, irreplaceable, and finite in number. They exist precisely once in the world because whatever is created in the future will be different—and centuries away. Ancient trees can perhaps be grown over two or three centuries, but such trees will not be Nature's trees in Nature's landscape; they will be humanity's trees in society's cultural landscape. Such trees will be different in the human mind, although they may be just as beautiful as those created by Nature. And, even if we start growing them today, we, our children, and our children's children for several generations will not be here to see them in old age—if they survive the unpredictable whims of wind, disease, fire, global warming, and finally human greed.

2. Old forests are living libraries in which we can study the biophysical principles of Nature's design elements that govern the world's forests and cities. As such, ancient forests are habitat to a number of organisms that either find their optimum habitat in them or require the structures provided by the live old trees, the large declining trees, the large standing dead trees, or the large fallen trees for parts of their life cycles. In other words, these forests are tremendous storehouses of biophysical designs that have been liberally borrowed to construct every city ever built. In addition, they are genetic reservoirs, harboring plants of potential use in designing new agricultural crops and medicines.

This has already proven to be the case in the forests of southern Mexico with their exceptionally rich biotic diversity, where in 1978 a wild variety of perennial maize (corn) was discovered. Through crossbreeding, this new strain could enable the corn-growing industry to avoid the seasonal costs of plowing and sowing. In addition, the wild germplasm offers resistance to several viruses that attack commercial corn. The economic benefits of this discovery could eventually reach billions of dollars.[18]

In North America, the Pacific yew, a shade-tolerant tree in the understory of old forests of the Pacific Northwest, was once considered a weed. Then, it became of medical importance because it was the sole source of Taxol for the treatment of ovarian cancer in women. Dr. Robert Schweitzer, president of the American Cancer Society, in a letter dated September 19, 1990, to Manuel Lujan, then secretary of the Department of the Interior, stated:

> On behalf of the two and one half million volunteers of the American Cancer Society, I write to urge that you take any and all actions to protect the Pacific Yew tree as a 'threatened species' pursuant to ... the Endangered Species Act. ... The American Cancer Society believes that you should designate the Pacific Yew ... as a threatened species. This action will ensure that women diagnosed with ovarian cancer in the years ahead will have access to a promising drug treatment.

The Pacific yew was once widely distributed from the southern tip of southeastern Alaska throughout the Pacific Coastal regions of British Columbia, Canada, southward to California. Today, however, the Pacific yew is rapidly disappearing because of clear-cut logging the old forests and because of uncontrolled commercial harvesting of its bark for medicine. But, as often happens in our commercialized world, once the pharmaceutical companies decoded the chemical substance of Taxol and could reproduce it synthetically, the Pacific yew was again relegated to treatment as a "weed."

3. Nature's biophysical principles function perfectly whether or not we understand them. In terms of Nature's biophysical principles, we would do well to learn more about their function from old forests and heed what we learn. To illustrate, each old tree is a "carbon sink," a storehouse of immobilized carbon, the storage of which reduces the potential carbon dioxide in the atmosphere and thereby has a positive influence on the greenhouse effect. There is, for example, 2.2 to 2.3 times as much carbon stored in a stand of 450-year-old Douglas-firs as there is in a sixty-year-old tree farm. In fact, landscapes with tree farms that are cut on rotations of fifty, seventy-five, and one hundred years would be able to store, at most, 38, 44, and 51 percent, respectively, the amount of carbon stored in an equally sized area of ancient trees. And, these values are conservative because the intense utilization of wood fiber in tree farms removes even more of the soil's capacity to store carbon.[19]

Although the reintroduction of trees into deforested regions will increase the storage of carbon in the living organisms of the area, conversion of old forests to young stands and tree farms under the current conditions of continual harvesting at forty to eighty years of age has added, and will continue to add, carbon dioxide to the atmosphere. These circumstances are likely to hold in most forests as long as the timber industry focuses first and foremost on maximizing profit—regardless of the environmental consequences such behavior engenders for all generations.

4. By saving these forests, we may be able to help future generations learn how to create sustainable forests, something no one has so far accomplished. These old forests are indigenous and so form a link to the past, to the historical forest. The historical view tells us what the present is built on, and past and present tell us on what the future can be projected. To lose the remaining old forests is to cast us adrift in a sea of almost total uncertainty with respect to the biological sustainability of the forests in the future because knowledge is only in the past tense, learning is only in the present tense, and prediction is only in the future tense. To have sustainable forests, we need to be able to know, to learn, and to predict. Without old forests, we eliminate learning, limit our knowledge, and greatly diminish our ability to predict how we could meet the requirements of future generations.

Because we did not design the forest, we do not have a blueprint, parts catalog, or maintenance manual to understand and repair it. Nor do we have a service department in which the necessary repairs can be made. Therefore, we cannot afford to liquidate the old forests that act as a blueprint, parts catalog, maintenance manual, and service department—our only hope of understanding the sustainability of the redesigned tree farm.

Moreover, we are playing "genetic roulette" with forests of the future. What if our genetic engineering, genetic cloning, genetic streamlining, and genetic simplifications run amuck, as they so often have around the world? Indigenous forests, whether old or young, are imperative because only they contain the entire genetic code for living, healthy, adaptable forests.

From intact segments of the old forest, we can learn to make the necessary adjustments in both our thinking and our subsequent course of management to help ensure the sustainability of the redesigned forest. If we choose not to deal with the heart of the issue of old forests, which is the biological sustainability of present and future forests, we will find that reality is more subtle than our understanding of it and that our "good intentions" will likely give bad results—and pass unpayable, ecological debts to future generations.

5. An old forest is a metaphorical mirror of what it means to be separate individuals *and* simultaneously a universal people. Whether we realize it or not, whether we admit it or not, we need one another. To illustrate, imagine the large, old trees of an ancient forest. Each signifies primeval majesty but only together do they represent an ancient forest. Yet, we do not even see the forest for the trees.

If we could see below ground, we would find gossamer threads of a special kind of fungus stretching for billions of miles through the soil. As explained in an earlier chapter, these fungi grow as symbionts on and in the feeder roots of the ancient trees. Not only do they acquire food in the form of plant sugars passed from the crown, down through the trunk and into the roots, but also, as conduits, they provide nutrients, vitamins, and water from the soil to the trees and produce growth regulators that benefit the trees. These symbiotic fungus-root structures (called *mycorrhizae,* in case you have forgotten) are the termini of the threads that form a complex fungal net under the entire forest and, as evidence indicates, connect all trees one to another.

Like the ancient trees, we are separate individuals, and like the ancient forest united by its belowground fungi, we are united by our humanity—our need for love, trust, respect, and unconditional acceptance of one another. Further, we must be able to share our *feelings* with at least one other person to find value in life. This suggests that, when all is said and done, we need one another because we grow out of the varied soils of culture and are thus united by the hidden threads of our common human needs. If, therefore, we lose sight of and touch with one another as human beings, we will find a diminishing value in life. And, our common bonds will progressively erode into ever-increasing fear and separateness.

Fear and separateness (which spawn destructive environmental competition and political wars) are choices made secretly in the human heart and acted out in the collective of society. Love, which fosters trust, respect, and mutual caring as part of a sustainable community, is also a choice of the human heart acted out in the collective of society. And, everyone has an equal choice, an equal vote, if you will. With my choice, I influence the politics of life by how I behave. With your choice, you do the same. Every choice counts, like every vote in a democracy, and in the end, the majority will rule.

Unless our minds and our hearts are set on maintaining an ecologically sustainable forest, each succeeding generation will have less than the preceding one, and their choices for survival will be equally diminished. While the choice is ours—and we are limited only by what we think we cannot do—the consequences belong to every generation from this point in time.

The Value of Saving Old Parts of a City

As the foregoing suggests, a forest, especially an old forest, is a classroom for learning about Nature's biophysical principles as functional elements of design. A city, on the other hand, is the practicum where we humans learn how to apply the biophysical principles and, through our application, to test our understanding of them.

1. Old cities, or parts of cities, are unique; once gone, they are irreplaceable—as well every archaeologist knows. That is why so much effort has been spent over many decades to find and excavate ancient cities of all cultures and preserve their detail as much as humanly possible in order to learn more about our collective story as we humans have journeyed through time.

How long will the mystery and the lessons enshrined in old cities and parts thereof be able to resist the onslaught of modern development that summarily razes the old to replace the thoughtful creation—the legacy—of bygone people and years with the imagineless, ticky-tacky creations of today's money chase? As with the vanishing old forest, some people feel a growing sense of impoverishment as historic buildings and areas of cities are allowed to fade unrequited into the unknown, forever lost to the impatience of today's hurry, worry society.

2. The few remaining centuries-old buildings marked by the passage of years with the accrual of age-related characteristics make them a noteworthy display of architectural design. In addition to being more creative, aesthetic, and daring architecturally, they are more closely aligned with Nature's biophysical principles in order to withstand the normal ravages of time.

The first architectural endeavor of a hominid was to find a suitable cave in which to live. (A hominid, *hom×i×nid*, is any of the modern or extinct primates that belong to the taxonomic family Hominidae, *Hom×in×idae*, of which we are members.) I say suitable cave because it had to be secure against cave bears and sabertooth tigers, among other predators. It had to offer sufficient protection against inclement weather and thereby retain heat for its inhabitants. It also had to be large enough to house a number of individuals and still be within a reasonable distance from water and a goodly supply of food. Survival dictated that each generation needed to pass to its offspring what it had learned by trial and error about the important qualities of a good cave. So, the rudiments of language came into being through an urge to communicate, to share with one another life's inner, private experiences of thought—the fertile bed of germinating ideas.

By two million years ago, an archaic species of human, known as *Homo erectus* ("erect man"), had developed the physical organs and mental capacity to string together three to five words at a time. This early language might have consisted mostly of simple nouns, the names of tangible objects, rather than abstract verbs and adjectives. By four hundred thousand years ago, the extinct early people of the Neanderthal Valley in Germany (*Homo neanderthalensis*—*ne×an×der×thal×ensis*—the suffix *ensis* means "belonging to") could do much better. So, a spoken, human language was gradually born.

Language guides thought, perception, sharing, and our sense of reality by archiving knowledge. Knowledge in turn is the storehouse of ideas, and language is the storehouse of knowledge. Language allows each succeeding generation to benefit from the knowledge gathered and compounded by generations already passed. It is a tool, a catalyst, a gift from adults to children. By means of language, each generation begins farther up the ladder of knowledge than the preceding one. In this sense, each building has, from the first cave inhabited by a hominid, been a functional text in humanity's architectural library of time. Each structure is thus a volume in the library, and language is the librarian that guards the knowledge harvested and stored through the ages.

Because ideas (like those leading to the chipping of a stone hand ax or the creation of a magnificent building) evolved over millennia with thought and language, they are part of our collective history in that ideas belong to everyone and are meant to be free. It is for this reason that old buildings and old parts of villages, towns, and cities are of such great value. They are the libraries of the past that teach of architectural foresight and creativity, such as the brave thrusting into the unknown world of consequences that created the first arch, the first dome, the first vaulted ceiling.

Clearly, pioneering creativity is beset with errors in understanding, interpretation, judgment, and application, but nothing is learned without errors. Nevertheless, it is errors that contemporary building codes were devised to correct, errors of structural weaknesses born of ignorance and the "shortcuts" of those who placed economic "efficiency" before structural "effectiveness," a point well made by Tammy Stehr, a local citizen:

> I keep hearing that Corvallis High School [in Corvallis, Oregon] should be torn down and replaced with new construction because it's old and worn out. This attitude, sadly, reflects the culmination of decades of mindless consumerism. It's the ultimate expression of a "throw-away society," where the old, the worn, the shabby is simply tossed rather than reused and recycled. ... When the car tires wear out, you replace them; you don't throw the car away. When the roof starts leaking, you repair or replace it; you don't bulldoze the whole house. When a button falls off your shirt, you sew it back on; you don't shred the entire garment.

> Even shiny new schools and houses often have mice, but you evict them and plug the ingress points; you don't sue the contractor. ... Wiring wears out, plumbing breaks down, door knobs fall off; responsible people get things repaired rather than throwing up their hands and expecting someone to hand them a replacement ... which ultimately will have the same problems, by the way! ...
>
> Buildings far older than CHS [Corvallis High School] (and less well-built!) have been responsibly and creatively rehabilitated across this country and elsewhere. (Their charms support a multi-billion dollar tourist industry in Europe and other regions.)
>
> Finally, as I understand it, school board members are actually stewards of the public resources the taxpayers have paid for. Wouldn't good stewardship involve maintaining buildings for the long term, rather than letting them deteriorate to the point that total replacement seems even remotely reasonable?[20]

Without retaining some of the old, however, the language of the new could not arise for there would be nothing concrete from which to learn, nothing to inspire alternative designs and thereby correct old design flaws. Moreover, there is an additional learning curve derived from old buildings. They are purposefully set ablaze and used by firefighters as a practicum, a concrete experience through which to learn the art of controlling and extinguishing unwanted fires that can threaten life and limb, home or business. Through this exercise, fire codes coincide with building codes in creating the safest possible structures. This also is part of the architectural library, and language is still the librarian guarding the knowledge stored there.

3. Nature's biophysical principles function perfectly in cities whether or not we understand them, as these few examples illustrate:

- Time and weather erode buildings, just as they do rock in creating soil.
- Gravity's tug is as irresistible in a city as it is in a forest.
- Termites and carpenter ants devour untreated wooden foundations of buildings the same as they do dead trees.
- Lightning strikes tall buildings as it does tall trees.
- Sewers overflow when clogged, just like streams and rivers.
- Ice storms break limbs and topple trees in both cities and forests.
- Tornadoes cut their destructive paths through cities as well as forests.
- Neither forest nor city is a stranger to fire; whereas fire's role in a forest is both destructive *and* creative in designing and maintaining a viable forest, in a city it is simply destructive, unless of course the building destroyed by fire was so poorly built that early destruction was inevitable anyway.

While these things may seem obvious, the outworking of Nature's biophysical principles is probably understood with greater clarity in forests, where they have long been studied, than in cities, where they have not. Although it is perhaps easier to accept their function in forests than in cities because we humans have less investment in the former, we are, in the end, no more in charge in one than we are in the other, something we are loathe to accept in either.

Given this, the abiding question is: How do we better *accept* and *apply* what we have learned about the systemic outworking of Nature's biophysical principles in forests to the design of cities, where such a great diversity of monetary capital is at stake, to say nothing of human lives? Here, I must point out that the question extends beyond the problem-solving scope of engineering; it requires the interdisciplinary, systemic thinking that crosses all bureaucratic, political, and discipline boundaries for the betterment of every city for all generations.

4. Cities archive knowledge of our cultural past we can read, but only if the old buildings remain, where they will become even more valuable for the education of future generations because they are continually recording our history. Each building is unique in that, like a book or a painting, it is a physical manifestation of the designer's insights as well as his or her knowledge of those biophysical principles that will allow the building to survive the rigors of time. In this sense, each old building is a test case that chronicles the effects visited on the structure by the stresses of weather and longevity that may, perchance, include the effects of an occasional earthquake, thereby compounding the building's value to current knowledge—the longer it stands.

5. An old building is a metaphorical mirror of what it means to be separate individuals *and* simultaneously a universal people. I say this because each person in a village, town, or city performs some kind of function that complements someone else's, that complements someone else's, that complements someone else's again, that forms the strands of an interdependent web of services that, in the collective, becomes "Everycity." The physical infrastructure is merely the fixed points that produce known landmarks around which we, the human web of service providers, move in making our appointed rounds as we tend to one another. With the possible exception of some crimes, I can think of no employment that does not serve someone, somewhere, somehow within the confines of a city. We are therefore individuals in our lives and universal in our living.

Summation

Chapter 9 visited age as an archive of history, wherein ancient trees and forests have much in common with historic buildings and old parts of cities.

Both have humble beginnings—a seed, a small building. Every living system, whether a forest or a city, sees its foundation become progressively older as the ever-newer components are added, and its beginning progressively passes into antiquity. To understand the story that both forest and city have to tell, we must begin at the beginning, something paleontologists and archeologists spend lifetimes trying to find by digging into the basement of time. And, what are they searching for—if not old buildings? These are old buildings of various ages that we, in today's insatiable, consumer-based, throw-away society are only too glad to be rid of in favor of something newer, shinier, more sensational, and disposable when something "better" (e.g., something newer, shinier, more sensational, and disposable) comes along.

Although Chapter 10 is not directly about the design interface between a forest and a city, it is indirectly connected because the type of economic system we embrace as a society will determine the degree to which we consciously plan our communities in harmony with Nature's biophysical principles—those that design the forest and nurture the city in the process. The more closely we emulate those principles, the more harmonious will be our lives and those of our children and our grandchildren.

Notes

1. T. S. Eliot. *The Rock.* The Rock: A Pageant Play. – London: Faber, 1934; New York: Harcourt, Brace, 1934
2. Georg Feuerstein, Subhash Kak, and David Frawley. The Vedas and perennial wisdom. *The Quest,* 8 (1995):32–39, 80–81.
3. Emma Crichton-Miller. Ecological drama. *Resurgence,* 223 (2004):26–27.
4. Robert Penn Warren. GoodReads. http://www.goodreads.com/author/quotes/3736.Robert_Penn_Warren (accessed January 22, 2009).
5. George Wilhelm Friedrich Hegel. http://www.wisdomquotes.com/cat_history.html (accessed January 22, 2009).
6. Robert D. Kaplan. Fort Leavenworth and the Eclipse of Nationhood. *The Atlantic Monthly,* 278 (1996):74–90.
7. William Blake. http://www.online-literature.com/blake/ (accessed January 23, 2009).
8. Colin Wilson. *The Atlas of Holy Places and Sacred Sites.* DK Publishing, New York (1996).
9. Chris Maser. Black bear damage to Douglas-fir in Oregon. *Murrelet,* 48 (1967):34–38.
10. The preceding discussion of architecture and history is based on James Howard Kunstler. Home from Nowhere. *The Atlantic Monthly,* 278 (1996):43–66.
11. Ibid.
12. Ibid.
13. Ibid.

14. Robert D. Kaplan. Fort Leavenworth and the eclipse of nationhood. *The Atlantic Monthly*, 278 (1996):74–90.
15. Sarah van Gelder. Beyond Greed and Scarcity. *YES! A Journal of Positive Futures*, (Spring 1997):34–39.
16. Ibid.
17. Caroline Myss. *Anatomy of the Sprit: The Seven Stages of Power and Healing*. Three Rivers Press, New York (1996).
18. L. R. Nault and W. R. Findley. Primitive relative offers new traits to improve corn. *Ohio Report*, 66 (November/December 1981):90–92.
19. Mark E. Harmon, William K. Ferrel, and Jerry F. Franklin. Effects on carbon storage of conversion for old-growth forests to young forests. *Science*, 247(1990):699–702; J. A. Kershaw, C. D. Oliver, and T. M. Hinckley. Effect of harvest of old-growth Douglas-fir stands and subsequent management on carbon dioxide levels in the atmosphere. *Journal of Sustainable Forestry*, 1 (1993):61–77; Joost Polak. Storing carbon and cleaning water: How to make profits without cutting trees. *Trendlines*, 1 (1999):4; Laurie A. Wayburn. From theory to practice: Increasing carbon stores through forest management. *Pacific Forests*, 2 (1999):1–2.
20. Tammy Stehr. Don't throw away good old buildings. *Corvallis Gazette-Times* (February 23, 2004).

10

Ecology and Economy

Ecology

> The most beautiful thing we can experience is the mysterious. It is the source of all true art and science. He to whom this emotion is a stranger, who can no longer pause to wonder and stand rapt in awe, is as good as dead: his eyes are closed.
>
> **Albert Einstein**[1]

Economy

> [American's] one primary and predominant objective is to cultivate and settle these prairies, forests, and vast waste lands. The striking and peculiar characteristic of American society is, that it is not so much a democracy as a huge commercial company for the discovery, cultivation, and capitalization of its enormous territory. ... The United States is primarily a commercial society ... and only secondarily a nation.
>
> **Emile Boutmy**[2]

To begin, it must be noted that we do not, in fact, destroy ecosystems as many people intimate, but we do alter them to various degrees. Nevertheless, they continue to function within Nature's biophysical parameters of cause and effect. In that sense, they function perfectly all of the time. What we do destroy is an ecosystem's ability to continue producing the products and services we valued it for in the first place. That is the ecological part of the issue.

The economic part of the issue—and the challenge for humanity—is to find an economic model (even if it is a modified form of capitalism) that fosters the kind of productivity in Nature that allows us to live in harmony with the ecosystems that nurture our very existence, our quality of life. To find an alternative view of today's cultural world, said social critic Ivan Illich, we have to learn an alternative language—something that works today in terms of the future. For that to happen, he said, we must return to the past to discover the history around which the current "certitudes" were invented—certitudes like "the ever-lasting truth of current knowledge," "positive growth," "the necessity of continual," "economic expansion," and "competitive necessity" because these form the organizational core of our modern experience, the bedrock of the reductionist mechanical worldview.[3]

According to a reductionist mechanical worldview, the economic process of producing and consuming material goods and services, has no deleterious effects on any ecosystem. This view is based on the *assumption* that natural resources are limitless, and that any unintended effects of the economic process, such as pollution and environmental degradation, are inconsequential. In contrast to this dominant worldview, however, the paradigm of sustainability is neither mechanical nor reversible; it is entropic, meaning the Earth's wealth and its ability to absorb and cleanse the waste produced by humanity's economic activities are both finite.[4]

In addition, "industrialism" is the name of our modern human economy, contended author Wendell Berry, that not only "thrives by undermining its own foundation" but also is "based squarely upon the principle of violence toward everything on which it depends, and it has not mattered whether the form of industrialism was communist or capitalist or whatever; the violence toward nature, human communities, traditional agriculture, ... local economies and land-use planning has been constant literally the world over."[5] Berry wondered if the economy can be fixed without radical change. I think not, but then neither can city planning be changed without a radical shift in the myopia of institutionalized economic thinking—and what we think dictates how we act and what we are as a culture.

> The Captains [and Priests] of Industry have always counseled the rest of us to be "realistic." Let us, therefore, be realistic. It is realistic to assume that the present economy would be just fine if only it would stop poisoning the air and water, or if only it would stop soil erosion, or if only it would stop degrading watersheds and forest ecosystems, or if only it would stop seducing children, or if only it would quit buying politicians, or if only it would give women and favoured minorities an equitable share of the loot? Realism, I think, is in a very limited programme, but it informs us at least that we should not look for birds' eggs in a cuckoo clock.[6]

By tying the economic process to the entropy of the physical world, economist Georgescu-Roegen pointed out that for Western industrialized society to survive with any semblance of dignity, it is critical to shift from the old reductionist mechanical worldview to a paradigm built around sustainability.[7] But, in making that point, he posed the unspoken question of how one measures sustainability in terms of human welfare. For the sake of discussion, three potential measures are highlighted: gross domestic product, eco-efficiency, and genuine economic indicators.

Before examining the gross domestic product and eco-efficiency, let me add that, according to Wendell Berry, people who use these two measures of economic viability are so far removed from the workings of Nature that they have no understanding of, feel no gratitude toward, and exercise no responsibility toward those systems that ultimately are responsible for feeding, clothing, and sheltering them.[8] Further, since economics is the language through which Western industrialized society was designed, it is the driving force wherewith we have created our current social-environmental dilemma of unchecked economic competition that has resulted in diminishing natural wealth, spiraling personal costs, rampant greed on the part of some of the most dysfunctional among us, and an increasingly uncertain future for all generations.

Gross Domestic Product

The motto of those who believe in the gross domestic product, as a measure of economic health, is, as I see it, to *make as much money as possible, any way possible, and pass the consequences to the next generation.*

The gross domestic product is nothing more than a measure of total output (the dollar value of goods and services) and tells very little in and of itself because it *assumes* that everything produced is, by definition, "goods," including people. William Bennett, who was President Reagan's secretary of education, observed that "socialism treats people as a cog in a machine of the state; capitalism tends to treat people as commodities." As such, the gross domestic product is an intellectual measure of the size of the U.S. economy, the amount of money that exchanges hands in a strictly additive sense, like an adding machine that cannot subtract and so makes no distinction between benefits and costs (credits and debits), productive and destructive endeavors, or sustainable and nonsustainable activities. In addition, there is no allowance for the declining quality of human life in the face of environmental degradation.

The reason for this disregard of human welfare is simply that gross domestic product treats everything that happens in the marketplace as a positive gain for humanity and thereby ignores everything that cannot be converted into money as unimportant to social well-being, such as the logging practices that decrease the late-season storage of snowpack that supplies cities with water. In this case, both logging and all the uses of water cause money to exchange hands and count as a plus in the valuation of the gross domestic product, even though degradation of the distant water catchments will eventually cause water shortages in cities. A prolonged shortage of water will put farmers out of business, a turn of events that will dramatically raise the cost of food and, consequently, increase the number of hungry people hidden in the United States under the shroud of the gross domestic product. Nevertheless, politicians generally see this decaying quality of human life through a well-worn, ideological lens that accepts economic growth as good—even as it cannibalizes the family, community, and environment that nurtures and sustains us.

On a more personal note, imagine yourself to be dying slowly of cancer, needing three major operations, while in the middle of a messy divorce that forces you to sell your home. You are an asset to the economy, from the gross domestic product point of view, because you are the cause of so much money exchanging hands.

In the first case, the farmer, as well as those with other water-dependent businesses, will be negatively affected, including everyone who in any way deals with food—from production to consumption. In the second case, your quality of life could hardly get much worse.

In both cases, the valuation of the gross domestic product goes up at the unmeasured expense of everyone's birthright to good-quality water and your losing everything you held dear to forces other than your impending death. This scenario is somewhat analogous to adding (crediting) the amount of each check you write against your bank account instead of subtracting (debiting) it.

The significance of this illogical calculation of economic activity revolves around the gross domestic product as the primary indicator of economic growth in the United States, the capitalist scorecard from one year to the next. As such, when growth in the gross domestic product exceeds 3 percent, it is usually favorable for incumbent politicians.

The danger hidden in the calculation of the gross domestic product as a real measure of economic growth is that it creates a false sense of prosperity and security, especially when growth is rapid because it ignores costs (adding only the benefits) and thus ignores the major problems confronting American society[9]—as depicted in Everyforest and Everycity:

- In a forest, calculation of the gross domestic product is like adding all of the inflowing cash from cutting and selling one's timber while ignoring both short-term costs (such as physical wear on roads and replacement of machinery and worn-out culverts) and long-term costs (such as the effects of soil compaction, siltation of streams, loss of soil fertility, fragmentation of habitats through clear-cut logging, the consequential restrictions of the Endangered Species Act, and so on).
- In a city, the calculation of the gross domestic product is like adding all of the inflowing cash from a shopping mall while ignoring both short-term costs (such as physical wear on buildings and equipment) and long-term costs of deferred maintenance (such as replacement of computer systems; resurfacing parking lots; replacing roofs; pollution of air, soil, and water; and so on).

Moreover, money itself as a measure of success is another example of seriously flawed thinking and valuation, as far as sustainability is concerned, because the bottom line in business is always stated as the truly important figure. The bottom line, showing how much profit has been made, is used as a measure of how well a company performs. Too little profit and a company is deemed inefficient, its management is slack, the full potential of its work-force is not harnessed, its products are out of date, or most damning of all, the company is not competitive in the global economy.[10] Are such damnations true? Is money the only valid measure?

Perhaps a family-owned furniture company is making products that are robust and lasting or selling its products to people with only a moderate income or those who are somehow disadvantaged. It may be paying higher wages to its employees than other furniture companies in the belief that all people deserve a living wage. It may be investing heavily in a strategy to protect the ecological integrity of its forestland wherefrom it draws the wood for its furniture or, having a noisy mill in a location being increasingly surrounded by people's homes, the company operates only one shift out of respect for the people living in the neighborhood.

In a world where money is the only acceptable measure of success, however, all these considerate gestures count as naught because traditional economists assure us that a linear notion of progress, meaning full steam ahead in the strictly material realm, is always the right course of action, whereas ecology is a discipline that teaches us the folly of speeding blindly into the future. In the scenario of full steam ahead, the quality of the products and the welfare of the people and the environment are all irrelevant in the face of a bottom line that is not performing as desired. The irony is that the bottom line profit actually accounts for only the last 10 percent of the total income earned, whereas the 90 percent of the monies that have been paid to earn the 10 percent are overruled and overshadowed by the 10 percent bottom-line profit.[11]

This type of valuation clearly points out that market economics places value on that which is deemed materially scarce instead of placing value on the worth of people and their potential for being loving, caring, honest, just, and thoughtful persons and neighbors. If we are to keep the softer social capital of mutual caring from becoming scarce, we must nurture it and reward it. But, this is one of the many areas in which the last 10 percent of the dollars, squeezed into profit margins at the expense of the 90 percent along the way, is simply not effective in meeting human welfare because it neither builds families and communities nor tackles poverty and protection of the environment.

Gross Domestic Product in Everyforest

An indigenous old forest has three prominent characteristics: large live trees, large standing dead trees or snags, and large fallen trees. The large snags and the large fallen trees are only altered states of the live old trees that become part of the forest floor and are eventually incorporated into the forest soil, where myriad organisms and processes make the nutrients stored in the decomposing wood available to the living trees. Further, the changing habitats of the decomposing wood encourage nitrogen fixation to take place by free-living bacteria. (As you may remember, nitrogen fixation is the conversion of atmospheric nitrogen to ammonia, a form usable by living organisms, such as a tree.) These processes are all part of Nature's rollover accounting system that includes such assets as large dead trees, biological diversity, genetic diversity, and functional diversity, all of which count as both *investments* (the sun's energy) and *reinvestments* (nutrients from the soil) of biological capital in the growing forest.

Current intensive, short-term, fiber-farm management disallows both the investment *and* reinvestment of biological capital (organic material, primarily decomposing woody debris) in the soil and so in the forests of the future because such capital has come to be erroneously seen as economic waste. For this reason, we in Western industrialized society plan the total exploitation of any part of the ecosystem wherefore we see a human use, and we plan the elimination of any part of the ecosystem for which we cannot see such a use. With this myopic view, we have created the intellectual extinction of Nature's diversity through our social planning system, an ill-advised point of view that will inevitably lead to the biological extinction of species and their functions within the global ecosystem, of which the world's forests are a part.

After the indigenous forest is liquidated, we may be deceived by the apparently successful growth of a first tree farm as it lives off the stored, available nutrients and processes embodied in the soil of the liquidated indigenous forest. But, without balancing biological withdrawals, investments, and reinvestments, biological interest and principal are both spent, thus committing biological and economic productivity to eventual decline. The dysfunctional "managed forests" (which are really "tree farms") of Europe—biological deserts compared to their original forests—bear testimony to such shortsighted economic planning.[12]

Converting a mature forest into a monocultural fiber farm or applying fertilizer to an existing monocultural fiber farm are neither biological reinvestments nor economic reinvestments in either the forest or the soil; they are economic investments in "crop trees." The initial outlay of economic capital required to liquidate the inherited forest, plant seedlings on barren land, and fertilize the young stand is an economic investment in the intended product. But, a forest does not function on economic capital. It functions on biological capital, bringing into question the economic practice of salvage logging.

In the mainstream of today's consciousness in the forestry profession, salvage logging to capture perceived economic waste in the forest is seen as a viable, even necessary, practice to maintain the flow of wood fiber from dwindling forests, especially old forests. Yet, there are two false assumptions in the industrial concept of waste: (1) To be of value, any potential forest product (such as a tree with usable wood) must in fact be used by humans or it is wasted, and (2) because any commercial product not used by humans is a waste, renewable natural resources carried into the future are economically discounted—negating the rights of the future generations in favor of immediate economic gain.

In contraposition, the biophysical sustainability of our forests depends on our ability to see the forest as a functional whole, by which I mean the ability to see commercial forest products in terms of a biologically sustainable system and a biologically sustainable system in terms of its ecological products. Unfortunately, we usually juxtapose product points of view and systemic points of view into a category of *either* products or forest health—instead of products *and* forest health—when asking questions of value. We then try to force science to answer the questions.

But, science, a discipline theoretically free to pursue biophysical knowledge for its own sake, is the language of the intellect and is concerned with how and why the universe works as it does. Scientific inquiry is not designed to deal with social values as they are the language of the heart. In contrast, economics is also a language of the intellect, but one that relegates the language of the heart into presumed monetary value.

While all three languages are necessary if human cultures and their manifold environments are to be mutually sustainable, the salient point is the state of our ignorance—not our intellectual illusion of definitive knowledge. And, it is exactly because we are so certain of our knowledge that we are often so abysmally unaware of our ignorance and the long-term effects of it, which we pass as our legacy to every subsequent generation.

To view the fatal flaws inherent in the tenets of the gross domestic product, read the current economic tenets of forestry that not only allow but also encourage people to spend the inherited forests of the world as though there were no tomorrow and to pass the bill forward to the generations of the future. This system would function thusly:

1. Promote and recognize the timber industry as the only industry that has the right and privilege of converting timber and nontimber products into financial capital regardless of the long-term consequences to the environment.
2. Annually clear-cut as much timber, primarily old trees, as one can sell.
3. Measure prosperity by economic activity and success by automation that eliminates people's jobs while increasing the profit margin.
4. Measure progress as the continual advancement of technology in the utilization of wood fiber from ever-younger trees.
5. Promote personal and corporate self-interest.
6. Encourage clear-cutting entire forests.
7. Take everything merchantable to the mill, even if nothing is left as a reinvestment of biological capital in the soil.

8. Erode and ultimately destroy biological, genetic, and functional diversity through centralized corporate economic competition that converts as much of the world's forests as humanly possible into quick monetary profits through monocultures of cloned and genetically modified fiber farms.

Gross Domestic Product in Everycity

Gross domestic product in a city can be summed up in two words—urban sprawl. To illustrate, 160 acres (65 hectares) of farmland, much of it prime farmland, is lost every hour (which equates to 2.7 acres or 1 hectare every minute) in the United States to urban sprawl, which means that half of the cropland in California will be gone in twenty more years. The irony is that, while one-fourth of the land lost will become parking lots and roads, the number of vehicles to use those parking lots and roads is growing six times faster than the number of people.

For instance, the human population in the western United States was only 4.1 million strong in 1900, about 5.4 percent of the 76.1 million people in the United States. In 1999, however, the region had 61.2 million people, or 22.4 percent of the nation's total population of 272.7 million. Since 1960, people have flocked to the Pacific Northwest, attracted by its quality of life, including a largely unspoiled environment. This influx caused the populations of Oregon, Washington, and Idaho to double, reaching 3,316,000, 5,756,000, and 1,252,000 people, respectively. In other words, the Pacific Northwest has grown as much in the past 40 years as it did in the first 157 years since 1803 when Thomas Jefferson purchased the Louisiana Territory.[13]

"When something keeps happening that no one much likes [such as uncontrolled growth in population and the resultant urban sprawl], and it happens in many different places," wrote Donella Meadows (now deceased), former adjunct professor of environmental studies at Dartmouth College, "it goes on happening despite all kinds of measures intended to stop it."[14] Then, we do not have a simple problem with our policies. We have a dysfunctional system that can be fixed only by thinking so differently that we create a different system based on radically different values.

Meadows went on to say that she had seen only one contribution to the discussion of urban sprawl that attacks the problem by completely rethinking how we use land, which fits well the notion that the level of consciousness that causes a problem in the first place is not the level of consciousness that can fix it. Meadows was referring to a book, *Better, Not Bigger,* by Eben Fodor, a city planner from Eugene, Oregon.[15]

Fodor began his book with a quiz. I have modified the quiz to examine the notion of gross domestic product (which reduces everything to isolated components), eco-efficiency (which attempts, albeit feebly, to keep some things connected), and genuine economic indicators (which strive to connect isolated pieces into a functional whole and keep them there) as they apply to a city.

As part of the quiz, bear in mind that the remaining fraction of relatively unspoiled land—around which swirls the controversy of urban sprawl—is the most precious asset of the northwestern United States, in addition to being one of the very things that draws most of the people to the area. Yet, once in the area, the new people contribute to the very urban sprawl that continually diminishes the quality and quantity of unspoiled land that drew them in the first place—a worldwide circumstance. With this in mind, consider the following questions in the context of gross domestic product:

1. How long do you want the buildings of your community to stand?

   "I am a contractor and I make my living by constructing buildings as fast as possible and as cheaply as possible, regardless of the selling price; so I build nothing to last more than thirty years."

   "I am a business owner and will expand my business or move it elsewhere, so I want the cheapest building with the least cost so I can leave it behind without a great financial burden."

2. Do you want new developments to be primarily mixed-use areas that eliminate, as much as possible, the need for automobiles, or are you content with housing developments that are separated from shopping malls and other necessary areas of town?

   "I don't want any mixed-use areas because the people living there might become too independent from the shopping mall and gas station I own, and giving those people an opportunity to meet their daily needs locally would cut into my profits."

3. Do you favor giving people a real choice in the kind of development that takes place within the city limits by forming a citizen's visioning process of the people, by the people, and for the people, or do you want to continue with the chamber of commerce holding sway in the effort to bring in more industry?

   "I want the chamber of commerce to lead the charge because continual growth is good for my business. Besides, the citizens are too extreme and emotional and don't know what's really good for the community in these uncertain times."

4. Do you want to limit new construction to resident developers and contractors who can be held accountable for what they build, or are you willing to accept absentee developers and their contractors, who build what they want and simply leave you with the consequences?

   "It's a free country, so I ought to be able to purchase land wherever I wish and develop it as I see fit, despite what local residents say they want. Besides, I don't think we have the right to limit anyone's ability to make money as long as they do it legally, under the letter of the law."

5. How much more traffic congestion do you want in your community?

   "I'd love a whole lot more traffic because I'm a car salesperson and my income depends on increasing traffic."

6. How much more pollution of the air, water, and soil do you want?

   "I don't want any restrictions because I am a salesperson for a large chemical company."

   "I don't want any restrictions because I am a farmer, and I need the chemicals to boost the production of my farm to its maximum."

   "I don't want any restrictions because I am a forester with a tree farm, and I want to control the unwanted vegetation so my trees can grow as fast as possible."

7. How much more prime farmland, timberland, and open space do you want to have developed?

   "I want to see all the land developed because I am a developer, and to me any land that is not developed to its fullest potential is just an economic waste."

   "I'm a farmer, and I want the right to sell my land for as much money as I can get so I can retire in luxury."

   "I own forestland, and I want the right to sell my land for as much money as I can get so I, too, can retire in luxury."

8. How much more of your local natural resources (open spaces, freshwater, electrical power, forests, grasslands, wildlife) do you want to consume?

   "We must, of necessity, sell all of our natural resources so we can look out for ourselves by competing effectively in the global market."

9. Do you want your city and county governments to continue subsidizing new developments, including industries that pollute your air, water, and soil, or should they use the money to purchase and protect open space, fund schools, offer day care for your children at community centers, create cultural and recreational programs to teach people about one another and foster interracial harmony, and perhaps still have money left over for a tax cut?

   "I am one of the 25 percent of the people whose business depends on development; I don't want anything to dampen my ability to earn all I can."

10. How much bigger do you want your community to get?

    "While I love the surrounding landscape and all, I want the amenities that big cities offer; I don't see why I can't have both. But I don't really want to live in a big city; I just want what they offer."[16]

Although this quiz blatantly points out the obvious negative effects of urban sprawl, we never hear about them because they attack the economic myth that, for an economy to be healthy, it has to be ever-expanding, and thus growth—all growth—is at once good for all of us all of the time and is deemed the very life's blood of economic health. If growth slows or ceases, our economy will indubitably collapse; that is the undying, fallacious message of the gross domestic product.

Clearly, therefore, the gross domestic product, with its myopic focus on dollars and its flawed logic, cannot be a measure of sustainability as it relates to human welfare. If not gross domestic product, then what could speak for human welfare? Many industrial participants of the 1992 Earth Summit in Rio de Janeiro, Brazil, touted a strategy of eco-efficiency that would refit the machines of industry with cleaner, faster, and quieter engines even as it allowed unobstructed prosperity, simultaneously protecting both economic and corporate structures.

Eco-Efficiency

The motto of those who believe in the gross domestic product as a measure of economic health is, as I see it, to *make money as efficiently as possible while protecting the environment as much as practicable.*

Industrialists hoped that eco-efficiency would transform the economic process from one that takes, makes, and wastes into a system that integrates economic, environmental, and ethical concerns. Here, you might ask what this notion of eco-efficiency is that industrialists around the world herald as their chosen strategy for change.[17]

Eco-efficiency is a term that primarily means doing more with less, a precept that Henry Ford was adamant about when he wrote in 1926, "You must get the most out of the power, out of the material, and out of the time."[18] His lean-and-clean operating policies saved his company money by recycling and reusing materials, reducing the use of natural resources, minimizing packaging, and setting new standards of human labor with his time-saving assembly line.

Although eco-efficiency is a well-intentioned concept that looks good on the surface, it is still within the bowels of the reductionist mechanical worldview with its current overlay of economic expansionism and thus is little more than an illusion of change. Rather than focusing on a new way of thinking, such as how to *effectively* save the environment, industrialists once again attached their hope to *efficiency*—the swan song of the environment—by which, unconsciously perhaps, they have set themselves up to quietly, persistently, and completely commercialize the world. This is but saying that eco-efficiency, while it aspires to make the old worldview less destructive, languishes from the fatal flaws hidden within the embrace of such destructive practices in the first place.

Although eco-efficiency aspires to make the reductionist mechanical worldview more benign through reduction, reuse, and recycling, it does not stop these economically driven processes of exploitation, needless overproduction, acquisitiveness, and pollution. The real message of eco-efficiency is to restrict industry and slow or curtail growth—to put limitations on the creative and productive capacity of humankind. Nevertheless, this message is simplistic because Nature itself is highly industrious, creative through unpredictable novelty, astonishingly productive, and even "wasteful" when viewed in the short term. The salient point is that Nature, unlike human industry, is *effective* in both the short and long term—*not efficient*.

Eco-Efficiency in Everyforest

If gross domestic product were refitted with the current notion of eco-efficiency, it would look something like this:

1. Recognize the timber industry as the only industry that has the right and privilege of converting timber into financial capital but *make concessions to nontimber interests* as long as they do not interfere significantly with cutting forests.
2. Annually clear-cut *fewer* acres and purposefully *hide* them.
3. Measure prosperity by *less* economic activity and success by introducing automation *more slowly*.
4. Promote *less-blatant* personal self-interest by *meeting or exceeding* many or most of the complex and often competing regulations.
5. *Encourage* saving a minimal buffer zone of nonmerchantable trees, but only along streams with anadromous fish, those that go to sea for part of their life cycles, like salmon.
6. *Encourage* leaving two nonmerchantable logs per acre as a reinvestment of biological capital in the soil.
7. *Standardize and homogenize* biological, genetic, and functional diversity by replacing forests with genetically engineered, cloned fiber farms for corporate economic benefit.

Eco-Efficiency in Everycity

Again, I modified Fodor's quiz to examine the notion of eco-efficiency as it applies to a city and land use planning. As a context for the quiz, know that the remaining fraction of relatively unspoiled land—around which swirls the controversy of urban sprawl—is at once the most precious asset of the northwestern United States and one of the very things that draws most of the people to the area in the first place. But, once in the area, the new people contribute to the very urban sprawl that continually diminishes the quality and quantity of unspoiled land that drew them to the community in the first place.

1. How long do you want the buildings of your community to stand?

   "I think fifty years would be long enough because too much in them will become obsolete."

2. Do you want new developments to be primarily mixed-use areas that eliminate, as much as possible, the need for automobiles, or are you content with housing developments that are separated from shopping malls and other necessary areas of town?

   "I wouldn't mind some mixed-use areas, but only if they don't interfere too much with my business in a shopping mall."

3. Do you favor giving people a real choice in the kind of development that takes place within the city limits by forming a citizen's visioning process of the people, by the people, and for the people, or do you want to continue with the chamber of commerce holding sway in the effort to bring in more industry?

   "I would prefer to have the citizens voice their opinion provided the outcome is still in favor of growth and business."

4. Do you want to limit new construction to resident developers and contractors who can be held accountable for what they build, or are you willing to accept absentee developers and their contractors who build what they want and simply leave you with the consequences?

   "I think resident developers and contractors would be the best choice, but I'm open to absentee developers and contractors as long as they follow the letter of the law."

5. How much more traffic congestion do you want in your community?

   "I suppose I have to tolerate a little more in the name of 'progress.'"

6. How much more pollution of the air, water, and soil do you want?

   "I suppose I have to tolerate a little more to have what I'm accustomed to enjoying."

7. How much more prime farmland, timberland, and open space do you want to have developed?

   "I suppose we have to sacrifice more land in the name of 'progress.'"

8. How much more of your local natural resources (open spaces, freshwater, electrical power, forests, grasslands, wildlife) do you want to consume?

   "I suppose we have to sacrifice some of our natural resources to create prosperity."

9. Do you want your city and county governments to continue subsidizing new developments, including industries that pollute your air, water, and soil, or should they use the money to purchase and protect open space, fund schools, offer day care for your children at community centers, create cultural and recreational programs to teach people about one another and foster racial tolerance, and perhaps still have money left over for a tax cut?

   "Well, I would be willing to accept a smaller subsidy in order to locate my business in such a quality environment."

10. How much bigger do you want your community to get?

    "I suppose we have to keep growing some because our city, county, and state governments say we do, which means whatever happens, will happen—but then, that's progress. In addition, we have no right to limit where people can live if that's where they want to be."[19]

Although eco-efficiency is marginally better than the gross domestic product, it still falls far short of measuring both human productivity and human welfare in any semblance of a commensurate manner. To accomplish this, we need to convert the current measurement of our economic system to one that, in fact, accounts for both credits and debits—a system of genuine economic indicators.

Genuine Economic Indicators

The motto of those who believe in the genuine economic indicators as a measure of economic health, is, as I see it: *Our economic system needs to be "people friendly"—present and future—in order to be environmentally friendly.*

A functional system of genuine economic indicators is critical because it is the best (and perhaps the only) accurate way to balance our social values with our growing knowledge of how ecosystems function and the limitations their long-term integrity imposes on both the potential and actual sustainability of our activities.

Think, for example, of Nature's inherent ecosystem services. We can neither replace Nature's services nor live without them, and yet they are not on everyone's balance sheet, meaning they are inadvertently discounted and liquidated in the pursuit of resources whose value in the marketplace *are* recognized and accepted. Nature's inherent services, on the other hand, are effectively finite—precisely because they are irreplaceable.

With respect to human economies, however, Amory Lovins, director of the Rocky Mountain Institute and author, contended that they are "supposed to serve human ends—not the other way around. We forget at our peril that markets make a good servant, a bad master, and a worse religion." In this sense, continued Lovins, right livelihood is certainly conducive to more worthy and durable values, and accounting for and being responsible for the protection of Nature's inherent services tends toward right livelihood.[20]

"We hardly know what we are doing," said author Wendell Berry with respect to Nature's inherent services, which we inevitably lose when we destroy habitat, "because we don't know what we are undoing."[21] In other words, the health of Nature's inherent services is based on the way Nature actually works, not on the way we want it to work. "In fact," said Lovins, "you can make a good case that probably half the GDP [gross domestic product] is pure waste, spent either to pay for or to remedy the effects of waste that shouldn't have occurred in the first place. … That's why so many people have the sense that they're running harder to stay in the same place."[22]

The notion of a system of genuine economic indicators is especially valuable because it provides an accurate way to balance our social values with our growing knowledge of how ecosystems work and the limitations their long-term integrity imposes on both the potential and actual sustainability of our activities. But, there is a caution to the statement that, as always, was elegantly envisioned by Wendell Berry. Namely, the effort to establish a system of genuine economic indicators will inevitably fail to treat the cause of the problem and thus leave intact the effects of the system of genuine economic indicators, which (1) is not sufficiently broad, (2) becomes a specialized movement or cause, and (3) is not radical enough to produce the necessary results. This admonishment requires proponents of genuine economic indicators to lead by changing their own private and public behavior, not just forcing others to change policy.

The worst danger, contended Berry, may be that proponents of genuine economic indicators lose the clarity of interpretation with respect to their own language, fostering its own confusion about meaning and practice while proffering the language into the hands of its opponents and vice versa. In either case, the language is too often relegated to the narrow vision of institutionalized intellectuals on both sides who find it easier to snipe than to lead. Once we allow our language to mean anything that someone else wants it to mean, it becomes impossible to say what we mean and mean what we say.[23]

Without a broad enough platform and the semantic integrity of language, environmentalists will continue to be viewed and chided as radical, subversive, extremist, antibusiness, antigrowth, and un-American by businesspeople because environmentalists rate the ecological values embodied in saving such things as old forests, wetlands, and endangered species to be greater than those of economic growth. On the other hand, environmentalists often view businesspeople as necessarily evil, greedy, and myopic by nature. That said, most of the problem with the economic point of view espoused by businesspeople, according to Thomas Gladwin, director of the University of Michigan's corporate-environmental management program, is that business executives and managers often lack good cross-training in science, as evidenced by the fact that less than 1 percent of 1.2 million articles written by business professors include the words *pollution*, *air*, *water*, or *energy*.[24]

Genuine economic indicators, in contrast to both gross domestic product and eco-efficiency, is a measure of total economic activity that includes both benefits *and costs* (credits *and debits*).[25] In this way, the citizens of a community could measure the true value and economic well-being of continual growth over time by assigning a monetary value to nontraditional economic indicators, such as the ecological integrity of the landscape where the community rests. By assigning either a positive or a negative value (a credit or a debit) to each such indicator, the indicators and their respective values can be combined into a single genuine economic indicator for the social-environmental welfare of the community.

Genuine Economic Indicators in Everyforest

By assigning a positive or a negative economic value (a credit or a debit) to nontraditional economic indicators, the owner or manager of a parcel of forestland (be it private or public) could measure the true value and biophysical well-being of the forest over time. These genuine economic indicators might include the amount of organic material left on a logged site as a biological reinvestment into the soil, the amount of siltation in streams, the temperature of the water in streams, the amount of erosion from roads, and so on.

Genuine economic indicators would then serve as a baseline for all subsequent deliberations concerning sustainable forestry. Another function of genuine economic indicators in forestlands would be to add value to those resources and activities that have no value in terms of the gross domestic product, such as ecologically sound road construction, saving dead standing trees for wildlife, protecting a tree with a colony of honeybees in it, and so on. A system of genuine economic indicators is an important tool because most, if not all, activities in a forest are omitted from valuation within the context of traditional economic measures. This omission becomes readily apparent when an honest deliberation is made over the biophysical well-being of a forest in terms of traditional economics and the legacy it engenders for the future.[25]

By way of example, one study of alternative strategies for managing the mangrove forests of Bintuni Bay in Indonesia found that leaving the forests intact would be more productive than cutting them. When the nontimber uses of the mangrove forests (such as fisheries, locally used products, and control of soil erosion) were included in the calculation, the most economically profitable strategy was to retain the forests. Maintaining healthy mangrove forests yielded $4,800 per 2.5 acres (per hectare) annually over time, whereas cutting the forests would yield a one-time value of $3,600 per 2.5 acres (per hectare). Maintaining the forests would ensure continued local uses of the area worth $10 million per year and provide 70 percent of the local income, while protecting a fishery worth $25 million per year.[26]

Another way landowners can make money from their forests without focusing solely on the cutting of timber is to use their forests for sequestering carbon. In New South Wales, Australia, for instance, David Brand, executive general manager of the state forests, watched the demand for timber declining, and in that decline he saw an opportunity to sell environmental services to local and foreign power companies that were looking for ways to offset the carbon dioxide their generating plants were releasing into the air. What would he sell? He would sell the sequestration of carbon (called "carbon storage rights") in the trunks and root systems of the forests' trees, for which he soon had an agreement to plant 2,500 acres (1,012 hectares) of degraded pastureland in eucalyptus trees at $10 for each ton (907 kilograms) of carbon sequestered. In Japan early in 1999, Tokyo Electric Power Company signed a letter of intent to plant up to one hundred thousand acres of trees over the next decade. "We don't need to cut timber at all any more," Brand said. "Our forests are being driven completely by environmental values."[27]

Selling the rights to store carbon is a smart move because forests are increasingly recognized as a major factor in the reduction of carbon dioxide, a greenhouse gas implicated in the concern over global warming. Creating a market for this service requires three main ingredients:

1. Formulate a framework of policy and political support to establish a level playing field that defines the commodities ("carbon credits") to be traded and that implements a system of credits and crediting that reduces financial risk.
2. Create a foundation of interested customers who want to reduce carbon emissions; educate them about the conservation of forests and sustainable-forest management as avenues for effectively reducing the amount of atmospheric carbon dioxide and thereby mitigate emission from electricity-generating plants.
3. Build the supply by helping owners of private forestland understand the dynamics of the carbon cycle in their forests, how to increase carbon storage, and how they can enter the carbon market with high-quality domestic carbon credits to sell.

To this end, the Pacific Forest Trust, headquartered in Boonville, California, has analyzed carbon storage under four types of variable-retention silviculture and compared them with clear-cutting, in which no carbon is stored. Results of the analysis showed that an additional thirty-two million tons of carbon would be stored on a given site for over fifty years under variable-retention harvesting. The analysis was based on three structural principles to ensure the credibility of the resulting carbon credits: permanence, additionality, and verifiability.

The foundation of the analysis is the *permanence* whereby the carbon will be stored, meaning one must assume that the gains in stored carbon are permanent by using such tools as conservation easements that would simultaneously protect a forest from being converted to nonforested use and ensure that its management would be permanently altered to increase the storage of carbon. The last would ensure that changes in future ownership would not reduce the gains in carbon storage. In Costa Rica, for example, high-quality carbon credits are currently derived from permanently dedicated parks and permanently secured conservation easements.

Additionality is a newly coined term that means a forest landowner has to do something significant in addition to what is currently done to ensure the trees are increasing their storage of carbon, such as letting them grow for a notably longer period of time than was previously allowed before harvest. A conservation easement that makes permanent all changes in management goals above prevailing norms is another example of additionality. *Verifiability* is ensured, as much as possible, by using well-documented data on the forest type and state-of-the-art modeling based on decades of published scientific research, as well as an annual, third-party assessment as required by the conservation easement.[28]

Although selling carbon credits has the potential to help reduce atmospheric carbon dioxide as a greenhouse gas, we can no longer assume, as stated earlier, that the services Nature offers free for the taking are always going to be there because the consequences of our frequently unconscious actions affect Nature in many unforeseen and unpredictable ways. What we can be sure of, however, is that the loss of individual species and their habitats through the degradation and simplification of ecosystems impairs the ability of Nature to provide the services we need to survive with any semblance of human dignity and well-being. Losses are just that—irreversible and irreplaceable. To keep such things of value as Nature's inherent services, we need to shift our thinking to a paradigm of sustainability whereby we calculate the full costs of what we do—a system of genuine economic indicators.

If the current notion of eco-efficiency were refitted with genuine economic indicators, it would look something like this:

1. *Eliminate* clear-cutting except where ecologically necessary to create or maintain biophysical sustainability.
2. Measure prosperity by the *choices* saved and passed forward to the next generation and the *richness* of things from which that generation could choose (natural capital) that accompanies those choices.
3. Measure productivity by the *biophysical integrity and sustainable capacity* of the forestland.

4. Measure progress by the *consciousness* underlying the care rendered the forestland as a biological living trust determined by a system of genuine economic indicators.
5. *Integrate* aquatic habitats and riparian zones in the forestland as part of a seamless, interactive whole.
6. *Eliminate the notion of waste* by seeing everything in the forest as part of the recyclable reinvestment of biological capital that maintains forest integrity and productivity.
7. View the need for regulation as *failure* in forestland trusteeship.
8. *Honor and protect* biological, genetic, and functional diversity as the principal of the biological living trust in order to protect the productive capacity of a given forestland to provide a sustainable level of interest in terms of economic goods and services for present *and* future beneficiaries.

To achieve the kind of revolution in consciousness called for by the paradigm of sustainability, we would do well to heed an ancient Arab proverb as a point of departure: Each word we utter should have to pass through four gates before we say it. At the first gate, the keeper asks, "Is this true?" At the second gate, the keeper asks, "Is it necessary?" At the third gate, the keeper asks, "Is it kind?" At the fourth gate, the keeper asks, "Is this something I want to be remembered for?"

How might this fit into caring for a forest as a biological living trust? Each thought and action in caring for a forest must pass through four gates: At the first gate of forest caretaking, the trustee asks, "Is what I am about to do ecologically sound?" At the second gate, the trustee asks, "Is what I am about to do necessary to the biophysical integrity of the forest over time?" At the third gate, the trustee asks, "Is what I am about to do biophysically kind to the forest?" At the fourth gate, the trustee asks, "Is what I am about to do something I want to be remembered for?"[29]

Genuine Economic Indicators in Everycity

Looking once again at Fodor's quiz as I modified it to examine the notion of genuine economic indicators as they apply to cities and land-use planning, think about the following: The remaining fraction of relatively unspoiled land—around which swirls the controversy of urban sprawl—is the most precious asset of the northwestern United States and one of the very things that draws most of the people to the area in the first place. Once there, on the other hand, the new people contribute to the very urban sprawl that continually diminishes the quality and quantity of unspoiled land that drew them to the area in the first place.

1. How long do you want the buildings of your community to stand?

   "At least a century so those constructed in my time can stand as a legacy to the future, as well as saving resources that would need to be used should the buildings decay earlier."

2. Do you want new developments to be primarily mixed-use areas that eliminate, as much as possible, the need for automobiles, or are you content with housing developments that are separated from shopping malls and other necessary areas of town?

   "If I had my way, I would opt for as many mixed-use areas as possible because the air would be much cleaner, the neighborhoods quieter, the streets safer, and the residents might actually get to know one another."

3. Do you favor giving people a real choice in the kind of development that takes place within the city limits by forming a citizen's visioning process of the people, by the people, and for the people, or do you want to continue with the chamber of commerce holding sway in the effort to bring in more industry?

   "I want the citizen's visioning because I want the most democratic process possible so that we and our children, as citizens, can have a real voice in the future of our own community."

4. Do you want to limit new construction to resident developers and contractors who can be held accountable for what they build, or are you willing to accept absentee developers and their contractors who build what they want and simply leave you with the consequences?

   "I definitely want local developers and contractors to do the building because they are more likely to live with the consequences of their work and so would be more conscientious than someone who is absentee."

5. How much more traffic congestion do you want in your community?

   "There is too much already; I am losing my quality of life and that of my children."

6. How much more pollution of the air, water, and soil do you want?

   "There is too much already; it is negatively affecting my quality of life and that of my children."

7. How much more prime farmland, timberland, and open space do you want to have developed?

   "It is imperative to save all that we have left in order to maintain our quality of life, protect our supply of water, have good-quality food grown by local farmers, and have open places for my children to experience the natural world."

8. How much more of your local natural resources (open spaces, freshwater, electrical power, forests, grasslands, wildlife) do you want to consume?

   "I think it is imperative that we conserve our natural resources as the wealth of our community and use them as wisely as possible because they benefit me by maintaining my quality of life and also are part of my legacy to my children, who have the same right to a life of good quality as I do."

9. Do you want your city and county governments to continue subsidizing new developments, including industries that pollute your air, water, and soil, or should they use the money to purchase and protect open space, fund schools, offer day care for your children at community centers, create cultural and recreational programs to teach people about one another and foster racial tolerance, and perhaps still have money left over for a tax cut?

   "I'll take the expanded services in which I deeply believe and the tax cut, which I can certainly put to good use for my children's health and education."

10. How much bigger do you want your community to get?

    "It's already big enough; but if it absolutely must grow, it has to do so in a way that protects the qualities I love it for, such as the beautiful landscape that surrounds it and the easy access to uncrowded places for recreation and spiritual renewal."[30]

Now, let us use the quiz to evaluate the differences among *gross domestic product* (strictly quantity oriented, even at the total expense of quality), *eco-efficiency* (moderately better because is seeks to do less harm but is still quantity oriented), and *genuine progress indicators* (oriented toward finding a balance between quantity and quality that allows the former while protecting the latter):

- In all of the answers under the category of genuine progress indicators, the intangible values that account for "a good quality of life and human welfare" would be honored and accounted for. The people who espouse these values, however, are often dismissed as "environmentalists," "crackpots," "antibusiness," "antiprogress," "undemocratic," "extremist," or some other such derogatory appellation.
- On the other hand, answers under eco-efficiency grudgingly accept continual growth, development, degradation of environmental health, a declining quality of life, and human welfare as a necessary

price of economic prosperity. I say grudgingly because there is a budding recognition that less harm needs to be done.

- Finally, the answers under gross domestic product propose continual growth, development, degradation of environmental health, the quality of life, and human welfare as the necessary price of economic prosperity or says: "Full steam ahead! I'll take mine now, and the future can worry about the leftovers." Those who espouse these values talk about the "real world" of competition, global markets, profit margins, and the vital necessity of continual growth in population and economic expansion for the long-term economic health and stability of the community. People with this attitude subscribe to the calculations of the gross domestic product, which deftly hides the real and lasting costs of continual growth, continual development, and urban sprawl.

The level of consciousness that inspired the notion of gross domestic product, and even eco-efficiency, does not have to reign, as explained by Donella Meadows. In Oslo, Norway, the quality of life appears to be measured by genuine economic indicators and their cumulative scores. "Oslo rises halfway up the hills at the end of a fjord," says Meadows, "and then abruptly stops"[31] because of a huge public park where no private development is allowed. The park is full of trails, lakes, playgrounds, picnic tables, and scattered huts, where one can enjoy a hot drink in winter or a cold one in summer. Tram lines radiate from the city to a number of locations at the edge of the park, which allows a person to ride to the end of a line, ski or hike in a loop through the park to the end of another tram line, and ride home again.

The park forms a "no-nonsense urban growth boundary" that effectively forces development inward. Consequently, there are no derelict blocks in Oslo because space that is no longer useful for one purpose is converted to another, better use. Urban renewal is everywhere a continual process of seeking the best and wisest use for all available space, which is, after all, the essence of sustainable development.

Because most streets in the shopping district are designed for pedestrian use, with limited space to park, cars are effectively all but eliminated. In addition, the trams are cheap, frequent, and go everywhere. The result is a city that is quiet, clean, friendly, attractive, and thriving and is surrounded by ample open space for recreation, exercise, and spiritual renewal.

Can you say this about your community, town, or city? Would you like to be able to honestly say this about your own town? Can we make our respective towns fit a similar description? Of course we can—it is only a choice. There is a long list of things that can be done, which Eben Fodor has organized under two categories: taking the foot off the accelerator and applying the brake.[32] Both of these come under the valuation of a genuine economic indicator but are totally absent from the valuation of the gross domestic product and eco-efficiency.

The accelerator represents the widespread public subsidies to urban sprawl, which, according to Fodor, include:

1. Free or subsidized roads, sewer and water systems, schools, and so on, *instead* of charging developers fees that are high enough to ensure the taxes of present residents will not go up to pay for the public services provided to new residents.
2. Tax breaks, grants, free consulting services, and other enticements to attract new, private businesses. Here, one might ask why the public should be forced to subsidize new businesses, especially when they undermine existing businesses that are paying their own way to be part of the community.
3. Waiving environmental or land-use regulations that will degrade the quality of life for all residents for all time.
4. Federally funded road projects that not only allow but also encourage further congestion of traffic.

Accelerators of growth make current residents pay for new development through higher taxes; lower services; more noise; less open space; more pollution of the air, water, and soil; and more traffic congestion. There is no legal or moral reason to keep a foot on the accelerators of growth, although most economists argue that a community will be healthy only if its economy keeps expanding, meaning the economists subscribe to floorboarding the accelerators of the gross domestic product. Easing up on the accelerators might at least guarantee that new development pays its own way (eco-efficiency). Finally, genuine economic indicators include long-term protection of the environment, even at the expense of growth because one *cannot simultaneously maximize both* quality and quantity. One must take precedence over the other, and when such things as open space are gone—they are gone and so is the present quality of life.

In contrast, applying the brakes means setting absolute, nonnegotiable limits to growth and new development, something that can be done in the following ways:

1. Establishing nonnegotiable growth boundaries and green belt systems of open spaces.
2. Protecting farmlands, forestlands, grasslands, shorelines, and so forth through nonnegotiable zoning.
3. Establish spending restrictions on infrastructure. Why should a Wal-Mart, K-Mart, or any other large chain store simultaneously suck in traffic and export local capital to corporate coffers while forcing the

resident public to widen the road at public expense? Another possibility is to let the narrow road control the traffic.

4. Downzoning, when appropriate, rezones the land so it can be used less intensively.
5. Make public purchase of development rights.
6. Either limit the rate of growth *or* place absolute limits on the size of a community that is commensurate with the limitations of its natural resources as determined by the connectivity and ecological integrity of its surrounding environment.
7. Conduct an impact analysis of all actions to determine the degree of loss or gain in social and natural capital.
8. Conduct a fiscal impact analysis to reveal the *true costs* of development to a community.

On one hand, there are unethical and illegal reasons for not wanting to apply the brakes: (1) to protect a special privilege, (2) to discriminate against certain kinds of people, and (3) to take private property for corporate gain without fair compensation. These reasons are based on calculations of the gross domestic product, fostering competition that is both socially and environmentally destructive.

There are, however, ethical and legal reasons for wanting to limit growth and new development: (1) to protect water catchments, aquifers, open space, farmlands, forestlands, grasslands, shorelines, and so on over the long term; (2) to protect the quality of the environment as a whole, particularly the quality of the air, as well as the quality and quantity of the water and the soil for all generations; (3) to slow growth by directing it into places where public services can be efficiently *and* effectively delivered in a way that will allow a community to absorb expansion and envision the long-term ramifications in a wise and psychologically mature way; and (4) to protect the resident community from outgrowing its land base by protecting all its natural resources from failure due to overdevelopment.[33] All of these reasons argue for the evaluation of growth using genuine economic indicators.

The system of genuine economic indicators would then serve both as a baseline for all subsequent deliberations concerning the social-environmental sustainability of the community's growth and as a means of measuring the *social-environmental effectiveness* of the criteria employed in land use decisions and subsequent actions based on those criteria. Another function of genuine economic indicators would be to add value to those resources and activities that have no value in terms of the gross domestic product or eco-efficiency, such as taking real care of the community's children and elders.

Genuine economic indicators are an important tool because most, if not all, activities in social-environmental planning, from the long-term health of the environment to the real welfare of the citizens, are omitted from valuation within the context of traditional economic measures. This omission becomes clearly apparent when a local government deliberates over the economic strength of a community's tax base in terms of traditional economics and as a legacy for the future.

Reuniting Ecology and Economy

Ecology and economy have the same Greek root *oikos*, meaning "house." *Ecology* is the knowledge or understanding of the house, and *economy* is the management of the house—and it is the same house. A house divided against itself is doomed to collapse, whereas a house united can stand an eternity.

Although a house divided cannot long stand, it has been the continuing assumption of our society that if we manage the parts right, the whole will correct itself. Yet, while evidence to the contrary now comes from all directions, our systems of knowledge, governance, and management are still structured around this assumption. The question before us, therefore, is how to reunite ecology and economy. To begin, I can take seriously the motto I wrote for genuine economic indicators: "Our economic system needs to be 'people friendly'—present and future—in order to be environmentally friendly."

Here, the question is how we go about doing this. Do we ignore strands in the biophysical web of social-environmental sustainability, as it were, or must we begin to acknowledge the environmental realities, based on our best scientific understanding as well as our highest social principles, and rethink our dysfunctional behavior? To give you an idea of the areas in which we need to reevaluate our behavior, I discuss making our economic system people friendly and making our economic system environmentally friendly.

Making Our Economic System People Friendly

There are many things that can be done to make our economic system people friendly at home. Each is common sense if you put yourself on the receiving end in terms of how you would like to be treated.

Treat others, as you want to be treated. In this case, Mother Teresa has said it best:

> People are often unreasonable, irrational, and self-centered; forgive them anyway.
>
> If you are kind, people may accuse you of selfish, ulterior motives; be kind anyway.
>
> If you are successful, you will win some unfaithful friends and some genuine enemies; succeed anyway.
>
> If you are honest and sincere, people may deceive you; be honest and sincere anyway.

What you spend years creating, others could destroy overnight; create anyway.

If you find serenity and happiness, some may be jealous; be happy anyway.

The good you do today will often be forgotten; do good anyway.

Give the best you have, and it may never be enough; give your best anyway.

You see, in the final analysis, it is between you and God; it was never between you and them anyway.[34]

[And if you do not believe in God; it is good business anyway.]

Do the job for others the way you would want the job done for you. In a business sense, this counsel specifically means to put yourself in your customers' shoes and ask yourself what kind of service would entice you to become a loyal customer who would also refer your business to others, then treat your customers accordingly.

Manage the only thing you can manage—yourself. The only thing that you can really manage is yourself—your thoughts, your attitudes, and your behavior. These are the things others will judge you by because most people are judgmental. This being the case, it is to your advantage to be authentic.

Authentic means to be trustworthy and genuine—"what you see is what you get," so to speak. Authenticity is the harmony between what you think, say, and do and what you really feel, the motive in the deepest recesses of your heart. The adage "deeds speak louder than words" is true as far as it goes, but what is left unsaid is that "motives speak louder than deeds."[35] A person is authentic only when personal motives, words, and deeds are in harmony with his or her attitude. Otherwise, as Ralph Waldo Emerson observed: "Your attitude thunders so loudly that I can't hear what you say."[36]

A person's attitude is the visible part of the person's behavior, but the person's motive is often hidden from view. When one's visible behavior is out of harmony with the person's motive, that attitude points to a hidden agenda that disallows a relationship of mutual trust to grow, a prerequisite for a sound business ethic.

Become a psychologically mature adult. By psychologically mature adult, I mean a person who takes responsibility for how he or she treats others. But, first you have to learn to treat yourself well, as an individual, before you can treat others well because you can only give what you have inside. To this end, strive every day to do something better than you did it yesterday. This effort will gradually raise the level of your consciousness of cause and effect, which is critical to your business because all we humans do, *ever*, is practice relationships. Consequently, all business is a practice in human relationships—nothing more, nothing less—and you set the tone by example.

There are five cardinal principles to becoming a psychologically mature adult:

1. You have to genuinely care about the welfare of the people you serve because people do not care how much you know until they know how much you care about them.
2. You must be honest and transparent in dealing with your customers because a lasting business is built on honesty—the foundation of which is trust. Moreover, you are obliged to be the first to trust because trust is earned, not lightly given. This means you are required to be consistent in how you treat your customers; consistency equates to knowing what to expect, which in turn equates to your customer's ability to trust a given outcome when dealing with you.
3. You cannot convince anyone of anything. To convince you that I am right, I have first to convince you that you are wrong, whereby I steal your dignity, leaving you little recourse but to defend yourself. So, your customers must clearly see how they get their value out of the transaction before you get yours.
4. Create and define your own set of values and stick with them in all your dealings. If your values are based on common sense and the essence of what it means to be a human, your business should enjoy repeat customers and referrals.
5. When things run amuck, as some things inevitably will, ask yourself: What can I learn from this? What can I do about it? And then act boldly on the answers.

Do it right the first time. Most people in the industrialized West seem to have the attitude that "time is money" and so are in a hurry to act, often without thinking through the potential consequences of their actions on either people or the environment. Some years ago, while working with the Shinto Priests in Japan, I learned a valuable lesson. The priests spent much time discussing the pros and cons of various decisions as well as the potential outcomes of this choice and that. While the meetings seemed interminable and inefficient to me, when the priests came to consensus, they acted with their collective wisdom the first time because they had winnowed the possibilities through an informal, but very effective, risk analysis. This experience taught me that patience and delayed gratification are often more efficient than instant gratification because doing it right the first time is both effective *and* efficient since it precludes the necessity of having to do it over at the extra cost of time, labor, and capital.

Leadership. The best leadership in the world is that of a good example.

Always give a little more than you get. If you always give a little more than you get, people will know they can count on you to give them their money's worth and will have an incentive to return. Moreover, your legacy will be to leave the world a little better for having been here, and that is worth more to human welfare than all the money in the world.

Focus on one small correction at a time. As a ship's captain makes one small correction at a time while plying the open sea toward a distant land, so a captain of business is obliged to do the same. The farther we predict into the trackless future, the more conscious and clear we must be of our vision, goals, objectives, and data in order to achieve the ideal of our dreams because we need to make one small correction at a time as early in the game as possible.

Although it is a basic truth that our economic system has to become people friendly—present and future—*before* it will become environmentally friendly, there are some things that can be done simultaneously in the environmental arena.

Making Our Economic System Environmentally Friendly

Ultimately, how we treat our environment is a mirror image of how we treat one another and ourselves. By way of illustration, if I want to make your acquaintance, it is incumbent on me to determine how I need to behave to encourage you to reciprocate in a manner that I would like. If I am rude, you will most likely dismiss me without a second's thought. On the other hand, if I am gracious, you will, in all probability, accept my invitation and respond in kind.

It is the same with the productive capacity of the land. If we continue to force ourselves on the land and demand that it produce commodities in ways that defy the biophysical principles of Nature's governance, we may well be the authors of our own demise. By the same token, if we require something from the land, we need to ask ourselves how we have to treat the land in order for it to produce what we want. If we then treat the land accordingly, we have the greatest probability of achieving that which we deem necessary to our survival. In essence, our success or failure is determined by how we think and thus behave.

How people think. In the consciousness and choices of individual people living and acting collectively in a local community lies a great power to heal *or* to sicken the Earth that supports us all. Before we introduce such things as pesticides into our local environment, therefore, we need to learn how to ask broad questions about the possible effects of our actions beyond our local borders. As you read, be aware that none of our actions are, or can be, carried out in isolation from the rest of the world. We cannot divorce ourselves from the systemic way the world functions, much as we might like to do so. Everything we do has an effect that is beneficial to some things and detrimental to others. Moreover, each effect is the cause of another effect—albeit one that we may not intend to have happen. That said, the kind of effect we have often depends on how we think.

Thinking in cycles ultimately causes us to see our lives as a circular dance wherein certain basic and necessary patterns of use and renewal, of life and death, are repeated endlessly. *Cyclical thinking* is process oriented in nature and is the ethical basis of indigenous American spiritual thought, as exemplified by Black Elk:

> Everything the Power of the World does is done in a circle. The sky is round, and ... the earth is round like a ball, and so are all the stars. The wind, in its greatest power, whirls. Birds make their nests in circles, for theirs is the same religion as ours. The sun comes forth and goes down again in a circle. The moon does the same, and both are round. Even the seasons form a great circle in their changing, and always come back again to where they were. The life of a man is a circle from childhood to childhood, and so it is in everything where power moves.[37]

Those who think cyclically humbly accept the mysteries of the universe. They allow Nature to teach them, and Nature's reflective lessons of infinite universal relationship are intrinsically valuable—to use something for its own sake and then to be the source of its renewal is to see it as a "re-source." In the original sense of the word, *resource* was a reciprocal relationship between humanity and Earth, a circle of taking and giving and taking again. The very structure of the word—*re* and *source*—means reciprocal relationship, a cycle, to use something from the Earth and then to be the source of its renewal.

People who see life as a great circle see everything as interdependent and nothing as independent. In Nature, there is no such thing as an "independent variable." Everything in the universe is patterned by its interdependence on everything else, and it is the pattern of interdependence and change that forms a constant. This constant is the principle of creation, infinite becoming, and thus infinite novelty.

The cyclical vision is at once realistic and generous. Those who accept it recognize that in creation lies the essential principle of return: What is here will leave and will come again; what I have, I must someday give up. They see death as an integral and indispensable part of life, for death is but another becoming, a view beyond a horizon.

In contrast to cyclical thinking, which arises from a desire to be in a harmonious relationship with the universe, our Western *product-oriented thinking* is linear in nature and oriented almost strictly toward the control of Nature as well as the conversion of natural resources into economic commodities—into money, the god of Western materialism. We suffer from "affluenza."

Wendell Berry offered an interesting point with respect to these two patterns of thought. He believes that the older, cyclical pattern of thought is complete, whereas the newer, linear pattern of thought is incomplete.[38] In other words, we lost part of ourselves somewhere in the count of ages.

Processes are invariably cyclic, however, rising and falling, giving and taking, living and dying in space on ever-expanding ripples of time. Yet, linear vision places its emphasis only on the rising phase of the cycle: on production, expansion, possession, youth, and life. It does not provide for returns, idleness, contraction, giving, old age, and death.

Waste is thus a concept that can only be born from a vision of economic linearity. According to this notion, every human activity produces waste because every human activity is linear.

The concept of linearity is the doctrine of eternal progress that is allegedly to free us from Nature's dictates. Entranced by this concept, society discards old experiences as fast as new ones are encountered, always thinking that we never "repeat" the old mistakes. Yet, in reality, we repeat them continually but deny it in our blind drive for material progress and so never learn from history, even if that history is but a year or two old.

In our minds, we are on an endless voyage of discovery. To return is merely to come back to the used because progress means always exploiting the new and the innocent.

Characteristic of the product-oriented vision is the notion that anything is justifiable as long as and insofar as it is immediately and obviously good for something else. Product-oriented thinkers require everything to proceed directly, immediately, and obviously to its perceived value. What, we ask, is it good for? And, only if it proves to be immediately good for something are we ready to raise the question of its value: How much is it worth? By this we mean how much money it is worth. Clearly, we ask this question because if it can only be good for something else, then obviously it can only be worth something else. Nothing, according to our economic system, has intrinsic value, not even money itself.

Although linearity in economics became resolutely manifest with industrial specialization, it is not economics that is linear, rigid, and narrow of focus, but rather economists who think they somehow have the answers to the world's riddles, answers that have escaped theologians, scientists, and philosophers alike.

Nonetheless, and in keeping with today's Western-industrialized thinking, current dictionaries define *resource* in a strictly linear sense as the collective wealth of a country or its means of producing wealth—hence any property that can be converted into money. Because of its linearity, product-oriented thinking discounts intrinsic value in everything it touches, including human beings.

Thus, it is not surprising that the intrinsic value of Nature is still largely discounted in our own culture. The same can be said of the intrinsic value of human beings when our military capacity for the destruction of the "foreign enemy" takes magnitudes of precedence over the domestic welfare and tranquility of our citizenry. Where does this kind of thinking lead us when we envision one another and ourselves merely as "human resources," a notion that inevitably translates into "human commodities"?

We can begin by looking at the education of our "resource managers." As soon as we demand, in this lifeless, linear sense, that education serves some immediate purpose and that it be worth a predetermined amount, we strip education of its intrinsic value, and it becomes mere "training." Such is the traditional training of foresters, range conservationists, fishery biologists, game biologists, and planners—both city and county, all of whom are trained in the traditional schools of resource management that abound in the United States. Once we accept so specific a notion of utility, all life becomes subservient to its use; its value is drained of everything except its "specialized use," and imagination is relegated to the scrap heap. In turn, these patterns of thought determine the core of a society's culture.

This type of reasoning makes difficult times for students and old people. Living either before or after their time of greatest social utility, it robs them of their sense of purpose. It also explains why so many species, both nonhuman and "primitive" humans, such as Indians in the Amazon basin, are threatened with extinction. Product-oriented thinkers perceive any organism not contributing obviously and directly to the workings of the dominant linear economy as having little or no value.

Jules Henry, an anthropologist/psychiatrist, said we Americans are a driven people, and all our activities are related to our "drivenness." It is easy to agree with this because the linear pattern of human thought produces a culture like ours, wherein *economics of acquisition is the force* that drives the society, determines its mode of institutions, and relegates spirituality to the bottom rung of the societal ladder. On the other hand, the cyclical pattern of human thought produces a culture, like that of the indigenous peoples of the Americas prior to the invasion by Europeans, in which *spirituality is the force* that drives the society and determines the mode of its economics and institutions.

Culture is based on and organized by the dominant patterns of human thought. Through the cultural dynamics of human-land interactions, these patterns of thought determine the construct with which a given society designs its villages, towns, and cities; the care it takes of its land; and the patterns it creates on the landscape. A society's culture is the product of its dominant mode of thinking; thus, given two identical pieces of land, each culture would, within a century, produce a different design on the landscape as a result of the pattern of its thinking—the template of individual values expressed in the collective mirror, the land. Because land and people are inseparably one, people unite with the land through their culture, for better or worse.

As the social values determine the culture, so the culture is an expression of those values. For this reason, the care given the land by the people is the mirror image of the hidden forces in their social psyche. These secret thoughts ultimately express themselves and determine whether a particular society survives or becomes a closed chapter in the history books. And, history books are replete with such closed chapters as the great empires of Mesopotamia, Babylon, Egypt, Greece, and Rome, all of which destroyed their forests and the fertility of their topsoil with their linear thinking, insatiable drive for material wealth, and warlike nature. In view of this catalog of extinct civilizations, one might do well to ponder where contemporary society—as defined by the global market—is headed.

If the biophysical systems on which we rely for our sustenance are to remain viable, our linear, product-oriented pattern of thinking must be guided by an understanding of, respect for, and compliance with the cyclical patterns that govern Nature's interdependent relationships. I say this because the linear, product-oriented mode of thinking can produce massive amounts of commodities deemed necessary to the survival of humanity but at the expense of the environment. Cyclical thinking, on the other hand, deals well with the interdependent relationships of Nature but not with the linear mode of mass production. We must combine the two modes of thinking to produce what we require in a way that recognizes, honors, and complies with Nature's biophysical principles and thereby protects the productive capacity of the land for all generations, an aspiration that dictates humility must guide what we introduce into the environment.

What we introduce into the environment. What any culture introduces into its environment, and the attitude with which the introductions are made, is determined by the mythological view of its place in and of creation. That notwithstanding, for a society to function so that its human members can survive requires Nature's cycles to be maintained in such a way that they provide enough energy for society to use. If some of the cycles we alter begin to deviate too much from the evolutionary track Nature has established, we tend to introduce "corrections" in that we seek new sources of energy, nurture new varieties of plants, and invent new modes of production. Some will be appropriate, in an environmental sense; others will not. Nevertheless, once we introduce something into the environment, it is immediately out of our control.

As I said earlier, we introduce thoughts, practices, substances, and technologies, and we usually think of these in terms of development. Whatever we introduce into the environment determines how the environment responds to our presence and to our cultural necessities. Accordingly, it is to our social benefit to pay close attention to what we introduce.

Our management of the world's resources is predominantly to maximize the output of material products—putting into operation the "conversion potential." In so doing, we not only deplete the resource base but also produce unmanaged and unmanageable "by-products," often in the form of hazardous "wastes." But, in reality, there is no such thing as a by-product. There is only an unintended product that more often than not is both undesirable and often capable of altering how our biosphere functions in unforeseen ways.

DNA from genetically engineered corn is illustrative because it has shown up in samples of indigenous corn in four fields in the Sierra Norte de Oaxaca in southern Mexico. This finding is "particularly striking" said Ignacio Chapela and David Quist, researchers from the University of California, Berkeley, because Mexico, the birthplace of modern corn, has had a moratorium on genetically engineered corn since 1998.

The fact is that corn genetically engineered to resist herbicides or to produce their own insecticides threatens to reduce the variety of plants in that region of Mexico because this corn may be able to outcompete the indigenous species. "The probability is high," said Chapela and Quist, "that diversity is going to be crowded out by these genetic bullies." This type of unwanted genetic transference is termed *genetic pollution*.[39]

In addition, the herbicide resistance could jump into weedy relatives and create "superweeds" that are beyond control. Moreover, plants that have been genetically engineered to produce their own insecticide can have serious, deleterious effects on insects and microbes in the soil that would also affect indigenous plants. This possibility is particularly troublesome because it could damage the vast natural repository of Mexico's indigenous corn, whose "wild, ancestral genes" might one day be needed to reinvigorate commercial corn.[40]

When enough human-altered cycles break down simultaneously, prudence dictates that we call into question the logic of our social system itself. Such scrutiny is wise because what society thinks of as ecological "corrections" are really self-reinforcing feedback loops, the outcome of which is not necessarily in keeping with our desires, regardless of what we try to do. Human societies either transform themselves in a truly corrective sense—realigning themselves with the biophysical principles—or they vanish.

As we come to recognize the undesirable effects of some of our introductions, we must be innovative and daring and focus on controlling the type and amount of processes, substances, and technologies that we insert into an ecosystem to effect a particular outcome. With prudence in our decisions about what to introduce and how, we can have an environment of desirable quality that can still produce a good mix of products and services, but on an ecologically sustainable basis. Hence, we need to shift our thinking from *managing* for particular short-term products to *caretaking* for a desired, long-term, sustainable condition on the landscape, an overall desired outcome of our decisions and actions.

If, as part of that process, we ensure that all material introductions we make into the environment are biodegradable as food for organisms like bacteria, fungi, and insects, then our "waste" would be their nutriment. In addition, if we use solar energy and geothermal energy instead of fossil fuels and nuclear materials and if we recycle all nonrenewable resources in perpetuity, we will shift our pattern of thought from one that is environmentally exploitive to one that is environmentally friendly and sustainable.

This said, the twenty-first century must begin the era of balances, of cyclic-linear environmentally sound economics wherein the health and welfare of our home planet takes precedence over our self-centered, short-term, materialistic wants. We must understand and accept that it is the collective thoughts, practices, substances, and technologies we introduce into the environment that determine the way the landscape will respond to our presence over time. That notwithstanding, many industrialists, just like other people, are reluctant to change.

The "not in my backyard" (NIMBY) stance has become the default for most communities facing any and all proposed developments. Such development (say, a commercial array of solar panels) is fine, even desirable—provided it is in someone else's backyard and does not affect the aesthetics of mine.

Closed-loop technology. Some years ago, I sat on a citizen's advisory council for a pulp-and-paper mill. This particular company uses water from the river that constitutes most of my hometown's drinking water. So, I naturally posed a question to the mill's manager:

> "What would happen," I asked, "if I took your intake pipe and put it immediately below your effluent pipe?"
>
> "We'd have to shut down the mill," he replied.
>
> "Why?" I queried.
>
> "Because the water's too polluted to run the mill," he said.
>
> "And you expect me to drink it," I countered. "Why don't you have closed-loop technology as far as water is concerned," I continued, "that would allow you to operate your mill without polluting my drinking water or the air?"
>
> "Because it's too expensive," was the reply.
>
> "Are you saying that we, the people, will have to force you to install and use closed-loop technology, even though it's available?" I asked.
>
> "That's about the size of it," he replied.

In other words, while some type of closed-loop technology is available for various industries, most will have to be forced to install it and then use it. Nevertheless, there is perhaps no better way to begin making our economic system environmentally friendly at home than to immediately begin protecting the quality of the three most important components of our environment—the air, water, and soil. Applied, closed-loop technology would go a long way in affording that protection, meaning people would have to change their thinking.

Understanding and accepting change. Change is the continual flow of cause-and-effect relationships that fit precisely into one another in time and space. Each cause creates an effect that becomes the cause of another effect ad infinitum. Thus, change comes on many levels and in many dimensions (large and small, gradual and sudden), such as those we can control to some extent and those we cannot. Change is forever ongoing because all things have within them the seeds of becoming something else.

All dimensions of change are fluid and dynamic, flowing together as rivulets that flow together as streams that flow together as rivers that flow together into the sea, where all waters merge and become dimensionless only to form again in the great cycle of raindrops, ice crystals, and snowflakes. Change by its very nature is the creative process and is a constant in the universe.

Understanding change is a matter of consciousness of the effect caused by a thought and subsequent action; the more conscious we are, the more flexible is our thinking. Unconsciousness, in contrast, is the lack of understanding the relationship between a cause and its effect and the lack of discipline to achieve that understanding. Hence, the less conscious we are, the more dogmatic and closed-minded we tend to be in our thinking. Here, we would do well to take some lessons from the Europeans.

Europe is leading the way. While the environmental standards are being lowered in the United States, the European Union is leading the way through initiatives like (1) recycling vehicles, (2) tracking biotechnology, (3) recycling electronics and banning toxins, (4) testing chemicals, (5) mandating the employment of "green design," (6) using carbon labeling, and (7) using carbon as currency.

1. "As of July 2002, auto makers were responsible for the recovery and recycling of all new vehicles they put on the market."
2. In the summer of 2003, guidelines were enacted for tractability, labeling, and environmental safety with respect to genetically modified crops.
3. As of August 2004, member states of the European Union required "European electronics companies to pay for the collection and recycling of their products. By July 2006, all electronic makers selling to Europe had to stop using some of the most notorious toxic substances: lead, cadmium, mercury, hexavalent chromium [the potential carcinogen featured in the film *Erin Brockovich*], and the chemical flame-retardants PBDE and PBB."
4. As of 2006, REACH (Registration, Evaluation, and Authorization of Chemicals) became effective. REACH is a program that "requires manufacturers to conduct rigorous safety and environmental tests on 10,000 common chemicals and find alternatives for the most-toxic substances."
5. The proposed legislation for "green design" mandates guidelines whereby all products will be environmentally friendly by design—"from computers to detergent bottles"—was drafted at the request of members in the European Union and has a high probability of passage in some form. This legislation would ensure the energy efficiency of all products, use of recycled materials in their manufacture, and limit the emissions of greenhouse gases as well as other hazardous substances.
6. One way to make rapid cultural change is to make people aware of the carbon content ("carbon labeling") in the goods and services they purchase. Every manager in England was aware of the carbon implications of their fleet of vehicles within months of the mandatory carbon/kilometer labels on new cars. Taxation is now based on these figures, giving drivers incentives to opt for the most fuel-efficient vehicles. Why not give everything a carbon label, from commodities

and buildings to rail travel and air travel? Why not make carbon labels as common as bar codes?

7. Another way to change culture is to create a "carbon currency." Although exchanging carbon credits is already beginning to happen among large, commercial producers and users of energy, it needs to take place at the personal level as well. A report, *Carbon UK*, stated that, "carbon will be the currency of the coming age."[41]

What could be more fair than issuing each person an annual allocation of carbon, much like giving a person an annual monetary allowance. Then, should a person wish to exceed their allocation of carbon, they could simply purchase an unused portion of someone else's at the fair market value. There are many people in the nonindustrialized countries who would be more than happy to sell their excess carbon allocation.

We could do the same. It is, after all, simply a choice on the part of citizens. As reporter and columnist William Greider pointed out: "There is nothing inherent to the functional principles of capitalism that requires it to be that way [exploitative]; that's a value choice made by people who have power within the system." Greider went on to say:

> The industrial system [in the United States] needs a thorough redesign of products and processes, reducing the damage to as near zero as possible. Europe is a counterweight to where the United States is: It is way down the road on a lot of the legislation that will drive the industrial transformation, like the laws mandating that auto companies take back and recycle cars at the end of their useful life. Europeans are convinced that it has to be done, so now it has value.
>
> Look at it this way: The consumer isn't getting any benefit out of the fact that big brands ignore environmental or social values. … It takes a purposeful minority with enough guts and daring and smarts to believe this could be changed, to see a different future than all these people around us—all these powerful people, all these powerful institutions—and then set out to change it. That's the process of history.[42]

If we, in the United States, have the wisdom, courage, and national grace to change the way we do business, we could become a truly great nation, one whose foundation is fairness and compassion—instead of the bullying tactics so ruthlessly employed by those addicted to the social trance of the money chase.

If we have the wisdom, courage, and national grace to change the way we do business, we could become a nation of trustees, caretaking the children's future—instead of the squanders of their inheritance.

If we have the wisdom, courage, and national grace to change the way we do business, we could, as a nation, earn the trust and respect of the world—instead of the mistrust and foreboding we are currently amassing.

If we have the wisdom, courage, and national grace to change the way we do business, we could find the endless possibilities for positive change—and their long-term benefits. We would do well to take most seriously our responsibilities as adult trustees of Nature's bounty because, in the words of Ralph Waldo Emerson, "Nature encourages no looseness, pardons no errors."[43]

Summation

In Chapter 10, I explored the unity between ecology and economy as it pertains to Everyforest and Everycity. The crux of the issue is that ecology and economy have the same Greek root *oikos*, meaning "house." Ecology is the knowledge or understanding of the house, and economy is the management of the house—and it is the same house. Because of that, good ecology is good economics, and good economics is good ecology. Since economics is the language through which Western industrialized society was designed, it acts as the driving force with which we create—for better or worse—our social-environmental reality.

Furthermore, since Everyforest serves the citizens of Everycity with an abundance of the necessary products and amenities of life, sound logic would dictate that the citizen of the city would do his or her utmost to caretake that which nurtures them and their descendants. To accomplish such an elevation in consciousness requires us first to recognize that our economic system will have to be transformed into one that is people friendly *before* it will become environmentally friendly.

The gist of Part IV is the requisite elevation of consciousness necessary to heal ourselves by focusing on healing the reciprocal relationships among our villages, towns, cities, and the landscape that sustains them all—present *and* future. To that end, Chapter 11 is a discourse about moving toward social-environmental equality in our diverse world, a prerequisite of social-environmental planning as if the dignity of all people really mattered.

Notes

1. Albert Einstein. http://www.quoteworld.org/quotes/4101 (accessed January 23, 2009).
2. Rachel While Scheuering. *Shapers of the Great Debate on Conservation*. Greenwood Press, Westport, CT (2004).
3. Ivan Illich. The shadow our future throws. *Earth Ethics*, 1 (1990):3–5.

4. Nicholas Georgescu-Roegen. *The Entropy Law and the Economic Process*. Harvard University Press, Cambridge, MA (1971).
5. Wendell Berry. In distrust of movements. *Resurgence*, 198 (2000):14–16.
6. Ibid.
7. Georgescu-Roegen. *The Entropy Law.*
8. Berry. In distrust.
9. The preceding two discussions of the gross domestic product are based on Timothy R. Campbell. "Sustainable Public Policy: Its Meaning, History, and Application." A paper presented at the annual conference of the Community Development Society in Kansas City, July 19–22, 1998; Georgescu-Roegen. *The Entropy Law*; Clifford Cobb, Ted Halstead, and Johnathan Rowe. If the GDP is up, why is America down? *The Atlantic Monthly* (October 1995):59–60, 62–66; Jane Silberstein and Chris Maser. *Land-Use Planning for Sustainable Development*. Lewis Publishers, Boca Raton, FL (2000).
10. Peter Lang. Money as a measure. *Resurgence*, 192 (1999):30–31; and David Boyle. The new alchemists. *Resurgence*, 192 (1999):32–33.
11. Ibid. Peter Lang. Money as a measure. *Resurgence*, 192 (1999):30–31
12. For a discussion of sustainable forestry per se, see Chris Maser. *Sustainable Forestry: Philosophy, Science, and Economics*. St. Lucie Press, Boca Raton, FL (1994).
13. Larry Swisher. NW faces issues of growth. *Corvallis Gazette-Times* (January 7, 2000).
14. Donella Meadows. Stopping sprawl. *Resurgence*, 198 (2000):30–31.
15. Eben V. Fodor. *Better, Not Bigger*. New Society Publishers, Gabriola Island, BC, Canada (1999).
16. I modified this quiz from Fodor. *Better, Not Bigger*.
17. Peter Lang. Money as a measure; Boyle. The new alchemists.
18. William McDonough and Michael Braungart. The next industrial revolution. *Atlantic Monthly*, 282 (1998):82–92.
19. I modified this quiz from Fodor. *Better, Not Bigger*.
20. The preceding discussion of natural capitalism is based on an interview of Amory Lovins by Satish Kumar. Natural capitalism. *Resurgence*, 198 (2000):8–13. (Amory Lovins is the coauthor, with Paul Hawken and Hunter Lovins, of *Natural Capitalism*. Little Brown, New York, 1999.)
21. Berry. In distrust.
22. See Note 20.
23. The preceding two paragraphs are based on Berry. In distrust.
24. Washington to launch new master's program. *Albany (OR) Democrat-Herald, Corvallis (OR) Gazette-Times* (January 24, 1999).
25. The preceding discussion of the genuine progress indicator is based in general on David Orr. Speed. *Resurgence*, 192 (1999):16–20; McDonough and Braungart. The next industrial revolution, 82, 83–86, 88–90, 91; Campbell. "Sustainable Public Policy"; Janet N. Abramovitz. Learning to value nature's free services. *The Futurist*, 31 (1997):39–42; Gretchen C. Daily, Susan Alexander, Paul R. Ehrlich, Larry Goulder, Jane Lubchenco, and others. Ecosystem services: Benefits supplied to human societies by natural ecosystems. *Issues in Ecology*, 2 (1997):1–16.
26. Abramovitz. Learning to value.

27. Joost Polak. Storing carbon and cleaning water: How to make profits without cutting trees. *Trendlines*, 1 (1999):4.
28. The preceding three paragraphs regarding the Pacific Forest Trust study are based on Laurie A. Wayburn. From theory to practice: Increasing carbon stores through forest management. *Pacific Forests*, 2 (1999):1–2.
29. The preceding two paragraphs on the four gates are based on Chris Maser. *Our Forest Legacy: Today's Decisions, Tomorrow's Consequences*. Maisonneuve Press, Washington, DC (2005).
30. I modified this quiz from Fodor. *Better, Not Bigger*.
31. The discussion of Oslo is based on Meadows. Stopping sprawl.
32. Fodor. *Better, Not Bigger*.
33. The preceding discussion of accelerators and brakes is based on Fodor. *Better, Not Bigger*; Silberstein and Maser. *Land-Use Planning*.
34. Mother Teresa. http://prayerfoundation.org/mother_teresa_do_it_anyway.htm (accessed January 24, 2009).
35. Chris Maser. Authenticity in the forestry profession. *Journal of Forestry*, 89 (1991):22–24.
36. Ralph Waldo Emerson. http://www.worldofquotes.com/author/Ralph-Waldo-Emerson/1/index.html (accessed January 25, 2009).
37. John G. Neihardt. *Black Elk Speaks*. Bison Books, University of Nebraska Press, Lincoln (1961).
38. Wendell Berry. The road and the wheel. *Earth Ethics*, 1 (1990):8–9.
39. The preceding three paragraphs on by-products are based on David Quist and Ignacio H. Chapela. Transgenic DNA introgressed into traditional maize landraces in Oaxaca, Mexico. *Nature*, 414 (2001):541–543.
40. Anita Manning. Gene-altered DNA may be "polluting" corn. *USA Today* (November 29, 2001); Mark Stevenson. Mexicans angered by genetically modified corn. *Corvallis Gazette-Times* (December 30, 2001); Will Weissert. Genetically modified corn called a threat to Mexican varieties. *Corvallis Gazette-Times* (March 12, 2004); Percy Schmeiser. Theft of life. *Resurgence*, 223 (2004):9–11.
41. The discussion of the new ideas in Europe is based on Samuel Loewenberg. Old Europe's new ideas. *Sierra* (January/February 2004):40–43, 50; Anthony Turner. Healing the air. *Resurgence*, 224 (2004):10–11.
42. Paul Rauber. A new mobilization is just beginning. *Sierra* (January/February 2004):38–39.
43. Ralph Waldo Emerson. Historic Quotes and Proverbs Archive. http://www.worldofquotes.com/author/Ralph-Waldo-Emerson/1/index.html (accessed January 25, 2009).

Section IV

A Century for Healing

For us, the human animal, our intellectual understanding, both scientific and social, acts as the process that interweaves our decisions and subsequent actions, transmitted via language, through the biophysical principles to create the social-environmental outcome of our desires based on our sense of values.

> I don't know what your destiny will be, but one thing I know: the only ones among you who will be really happy are those who will have sought and found how to serve.
>
> **Albert Schweitzer***

* Albert Schweitzer. http://209.85.173.132/search?q=cache:FmNNynu8GgkJ:www.schools.utah.gov/curr/lifeskills/servicelearning/pdf/servicequote%2520cards.pdf+I+don't+know+what+your+destiny+will+be,+but+one+thing+I+know:+the+only+ones+among+you+who+will+be+really+happy+are+those+who+will+have+sought+and+found+how+to+serve.&hl=en&ct=clnk&cd=2&gl=us (accessed January 24, 2009).

11

Toward Social-Environmental Equality in a Diverse World

> Those who say it cannot be done should not interrupt the person doing it.
>
> **George Bernard Shaw**[1]

Paul F. deLespinasse, a professor emeritus of political science at Adrian College in Michigan, views the global movement of jobs, much like an ecologist might view the movement of soil. Erosion of soil is normally thought to be necessarily bad, just like the outsourcing of jobs, but whether it is depends entirely on your point of view.

The Beginnings of International Social Parity

Farmers, who see the topsoil wherein their crops grow washed away by water, would naturally think such erosion is bad because loss of the topsoil would affect their livelihood. But then, farmers think about an infinitesimal spot on the Earth's surface with which they are familiar and to which they have a sense of commitment. On the other hand, Nature moves soil around all the time through such agents as wind, water, ice, and gravity. In this way, soil in a garden may instantly move an inch or two under the sudden impact of falling rain, whereas soil from a mountainside may end up on the ocean's continental shelf as the centuries and millennia pass.

deLespinasse makes a similar point with respect to jobs moving from the United States to other countries. This "outsourcing," as it is called, threatens the job security of an increasing number of American workers, just as the loss of soil from a field does a farmer's "crop security." For example, a company that produces computer systems might disseminate the production of some high-end products to a number of countries: A product's development might be conceived in Singapore; approved in Houston, Texas; conceptually designed in Singapore; production designed (engineered) in Taiwan, as well as the initial manufacturing of the components; and finally assembled in Singapore, Australia, China, and India. Once assembled, those products made in Australia, China, and India are sold primarily in those markets, while products assembled in Singapore are marketed throughout Southeast

Asia. Predictably, big business is mounting a quiet offensive to quash any state or federal efforts to retain jobs in the United States or otherwise constrain economic globalization.

Even so, outsourcing is like soil being redistributed from one place to another. In this case, some of the material wealth of the industrialized West is finding its way eastward, where it is beginning to equalize the quality of life we think of as "standard of living" for people in poorer countries. In the United States, however, American workers see the redistribution of jobs as a threat, just as a farmer sees erosion of the soil. In both cases, there is a natural tendency to try protecting what we have from being dissipated to other areas.[2]

In keeping with the above notion of equalizing material wealth, former federal reserve chair Alan Greenspan warned against resorting to "protectionist cures" in order to stem the hemorrhaging of traditionally American jobs to other countries. Greenspan went on to say: "The protectionist cures being advanced to address these hardships will make matters worse, rather than better. Protectionism will do little to create jobs, and if foreigners retaliate, we will surely lose jobs."[3]

Although no farmer can protect all the soil in his or her field from Nature's agents of redistribution, each farmer, enlisting the aid of imagination, can build up the organic fertility of the soil in the field by employing wise farming practices. Likewise, no rich, industrialized nation can forever hoard all of the jobs that feed people when much of the world is starving. But, each industrialized nation can employ the imagination and inventive talents stored in the minds of its citizenry to create new jobs that help caretake the nation and its people rather than squandering its social resources on war, however it is defined. Here, the first line of defense against the loss of jobs is a good offense—investing in a truly excellent public educational system that goes beyond math, science, and engineering to include the humanities, all of them, to move the nation toward psychological wholeness and maturity for the sake of all generations.

Whether any of our political leaders accept this counsel, the material quality of life will improve for some people as jobs flow into nonindustrialized or moderately industrialized countries, assuming they are dissatisfied with what they currently have, which may not always be the case. As a result of this job redistribution, the rampant materialism in the United States will likely slow and, perhaps, come to a rather sudden halt. While this scenario will be most unwelcome, even downright frightening here in the United States, it is inevitable because no nation can forever parasitize the world for its exclusive benefit without eventually being forced to share the material wealth as the decades glide silently into history—witness Great Britain and its colonial empire.

Equalizing the material wealth of the world will take years but is necessary if humanity is to survive itself and the current, undisciplined, material appetites of the monetarily wealthy few. If we, in the United States, are to exercise the role of a true world leader with any sense of nobility, we must begin now to learn and teach our children that there are two ways to material wealth and a high-quality lifestyle—work even harder or want less. The latter is the more gracious of the two because it allows us to share our abundance. With the former, there is never enough to satisfy the fear of scarcity and hence a restlessness of the ego that manifests in an aggressive attachment to and recognition of the physical. With the latter there is a growing sufficiency and hence a contentment of the soul, which is manifested as a gentle, progressive detachment from the physical. It translates as a pervasive sense of well-being, as an economy of life and energy in the best sense of the words. Commensurate with this lesson is knowing when enough is, in fact, enough.

When is enough, enough? It seems like a simple question. When you have eaten your fill, for example, you quit eating because you know you have had enough, for the moment at least. But, what about enoughness in the material sense, other than being immediately satiated with food? Although often referred to as a "standard of living," our sense of material enoughness really has to do with our level of fear and our sense of survival. Unfortunately, for many people, the sense of survival is based on an escalating disaster mentality that constantly conjures the next more frightening calamity for which they do not think they have enough material things to feel secure. And, according to insurance companies, there is always some catastrophe on the horizon that we do not have enough insurance to cover unless, of course, we are already heavily insured against it, whatever "it" is.

This sensation of never having enough relies on the notion that once our immediate needs are met and fears quelled, our desires have to be aggrandized. The ever-increasing rates of acquisition of unnecessary products and their faster disposal feed the manufacturing industry first and garbage dumps, euphemistically called sanitary landfills, second.

The ever-expanding plethora of catalogs points out that our materialistic, social appetite seems to have reached a compulsive, addictive state wherein to want is to have to have. Ralph Waldo Emerson called this our "bloated nothingness."[4]

What a stark contrast our society is to the simplicity that was everywhere visible at Gandhi's ashram in India. As individuals, the people living at the ashram took the vow of *maganvadi,* of "nonpossession." We, on the other hand, have made synonyms of "desire, want, need, demand" and in so doing have lost sight of the land's productive capability as well as of the difference between life's necessities and materialistic desires—the illusions of happiness and our inner sense of leisure.

The Chinese character for leisure is composed of two elements; one means "open space" and the other "sunshine." Hence, an attitude of leisure creates an opening that allows the sunshine to enter. Conversely, the Chinese character for busy is also composed of two elements; in this case, one means "heart" and the other "killing." This character points out that for the beat of one's heart to be healthy, it must be leisurely.[5]

"Cluttered rooms and complicated schedules interfere with our ability to treasure the moment," wrote Victoria Moran in her book *Shelter for the Spirit*. "Ironically, our houses and apartments—where some of our best moments can be—seem to attract clutter and complication like a magnet."[6] Even if we lack the inherent ability for simplifying and culling the inconsequential from the basic necessities, we can all learn to do so. Such discernment is the way to intentionally create a home and a life that nurtures our spirit, a life that separates the essential from the nonessential—and leisure is essential.

We tend to think of leisure, according to Brother David Steindl-Rast (a Benedictine Monk), as the privilege of the well-to-do. "But leisure," said Brother Steindl-Rast, "is a virtue, not a luxury. Leisure is the virtue of those who take their time in order to give to each task as much time as it deserves. ... Giving and taking, play and work, meaning and purpose are perfectly balanced in leisure. We learn to live fully in the measure in which we learn to live leisurely,"[7] a sentiment echoed by Henry David Thoreau: "The really efficient laborer will be found not to crowd his day with work, but will saunter to his task surrounded by a wide halo of ease and leisure."[8]

In its turn, enoughness is the key that unlocks the door of contentment, that part of life that fills the soul to overflowing and more than compensates for less materiality. Contentment means having what you truly want, truly wanting what you have, and the wisdom to know the difference.

Before the heart of "contentment" can be truly discussed, it is necessary to understand and integrate two perspectives of time—that of a clock and that of an hourglass. Time, as measured by the ticking of a clock, is constant in tempo. With a clock, you see the hands move from second to second, minute to minute, and hour to hour—as 'round and 'round the clock's face they go. While to a youth time seems to drag, even stand still, to an older person it seems to fly despite the fact that watching a clock's hands make their appointed rounds belies both the impatience of youth and the sensation of time as fleeting in old age.

Contrariwise, if you measure time through the functioning of an hourglass, you have the distinct impression that time is "running out," like the sand pouring to the beck and call of gravity from the top of the hourglass, through the small hole in its middle, to the bottom. Most adults view time with a growing sense that theirs is running out, so they feel compelled to grab all of life they can before their time is "spent," a fear of loss that champions material acquisitiveness in the supposed safety of the status quo. This sense of impending loss as time runs out causes people to avoid, as best they can, the admission of bodily changes wrought by the inevitable advance of aging.

In reality, of course, time does not run out; our bodies expire instead. And, it is precisely the dual sense of time running out and the demise of our bodies that causes many people to forfeit contentment in a bid to stop time.

Contentment is the inner feeling of fulfillment, of having enough. It can be likened to a quiet hermitage from which one can gaze on the harried world and discover that it holds within its material clutches nothing one wants. This does not mean a desire never enters one's head because now and then it does. But, when a desire finds its way into one's consciousness, it calls for reflection on whether the object of desire will add in some way to the spiritual quality of life or only to its material clutter. This is a necessary question because life's material wants are restless, fleeting whims of an uncontrolled mind.

Whether we learn when enough *is* enough and thereby enjoy the contentment enoughness brings, the world of humanity is changing and will continue to do so. This change—international social parity—must someday force the industrialized West to revisit the notion of development.

Revisiting the Notion of Development

Of the several facets reflected in the term *development*, we in the United States have chosen to focus on a very narrow one: development as material growth through centralized industrialization, a notion we glibly equate with social "progress" and "economic health." The narrowness of this view is, I believe, behind the notion of "developed" versus "developing" nations.

I have over the years worked in a number of countries without giving much thought to the notion of developed versus developing or, as some would put it, "underdeveloped," although I have spent time in both. That said, I was profoundly struck by the arrogance and the narrowness of such thinking during a trip to Malaysia some years ago.

Malaysia is the only place where I have ever heard the people refer to their own country as developing, as though they are lesser than developed countries and must somehow catch up to be equal. Yet, the Malaysians had a national unity like none I had ever before seen, not even in the United States, where all my life I have been taught about and heard about human equality—an ideal I have not seen practiced.

Malaysia is as great a mixture of cultures, national origins, and religions occupying as small a space as I have ever experienced. Yet, when I asked people what their ethnic background was, they referred to themselves as Malaysian Chinese, Malaysian Indians, Malaysian Sri Lankans, and so on. But were I to ask such a question in the United States, the response would be African American, Chinese American, Japanese American, German American, Italian American, and so on. While the difference may be subtle, it is profound. The Malaysians focus on their unity, while we in the United States focus on our sense of separation.

On any given day in Malaysia, I ate breakfast the Malay way, using both hands, with a spoon in one and a fork in the other. At lunch, I ate with chopsticks, and at supper I ate as much of the world eats, with my right hand as the only utensil. There were even four hour-long evening news programs, one each in Malay, Indian, Chinese, and English. Of course there were social problems, but I have never before experienced such integration of differences into a sense of wholeness.

As a guest and stranger in Malaysia, I felt that sense of wholeness encompass me. I felt welcomed and accepted for what I was as a human—not who I was as a nationality. In a strange, indefinable way, I felt more at ease and at one with the people of Malaysia than anywhere I have ever been.

If this is not development, what is? But, then it depends on how one defines development. If development is defined as a certain material standard of living based on the consumerism of centralized industrialization, Malaysia is indeed behind the United States. But, if development is defined as social civility and tolerance, the United States, compared to Malaysia (a nation even younger than the United States), is a *developing* country.

And, what about aboriginal peoples who not only have civility and tolerance but also have a long-term sustainable relationship with their environment. Are they not developed?

It is ironic that the very people who deem themselves to be developed and so "civilized" are the ones who have so ruthlessly destroyed the cultures of those they unilaterally brand as "undeveloped" and therefore "uncivilized savages." Fortunately, despite the continuing onslaught of civilized peoples, there are a few remaining hunter-gatherers, some of whom live in the deserts of the Middle East, North Africa, and Australia, as well as the Arctic, the plains of southern Africa, and the jungles of South America and Papua New Guinea.

I say fortunately, albeit they are severely endangered, because there is much about development and sustainability that we in the industrialized world can *relearn* from them. After all, our ancestors were also indigenous, tribal people at one time. Our problem of late is that we have forgotten most, if not all, of the wisdom they once knew. And, it is precisely this loss of ancient wisdom that is forcing us to focus on a contemporary question.

How must we view development if the concept is to be equitable and sustainable? Make no mistake, all parties must view "develop," "developing," "developed," and "development" as equitable if development is ever to become sustainable.

In placing development within a new context of conscious choice, answers to the following questions will be very different from today: What do we mean by development? What do we mean by developing? What do we mean by underdeveloped? What do we mean by poverty? If a lifestyle promotes sustainability through conscious choice, conscious simplicity, self-provisioning, and recognizes the relationships between a person's sustenance and the livelihood of the person's immediate surroundings in relationship to the larger world (their fidelity to their sense of place), that life is not necessarily perceived as one of poverty. This leaves the way open to change the indicators of development.

Progress, in this sense, would be any action that moves a person, community, culture, or society toward social-environmental sustainability. For society to progress, decisions need to recognize and respect the requirements and rights of future generations as well as the requirements and intrinsic value of all species and the Earth's biophysical carrying capacity with respect to its human population. (*Carrying capacity* is the number of individuals who can live in and use a particular landscape without impairing its ability to function in an ecologically sustainable way for future generations.) This proposition is very different from our blind faith in the material progress we think of as development.

Again, I think the narrowness with which we view development (i.e., the centralized production of material consumer goods through industrialization) is one root of the arrogance exhibited by Western industrialized countries that designate themselves to be "first-world" nations and all the others as "second-" or "third-world" nations. In Canada, however, the aboriginal peoples have turned this notion around.

Although the Canadian government refers to the aboriginal peoples as "Indian bands," the people think and speak of themselves as "First Nations." The people think of themselves as First Nations because they were among the original people, the indigenous people on the land in time and space, long before any of the outside invaders from Europe even knew that the "New World" existed.

Add development to this time-space sense of First Nation, and a clearer picture emerges. The indigenous peoples, the first humans in what is now Canada, had developed a lifestyle that had long been sustainable in and with their environment, despite the fact that they warred among themselves. Yes, the invaders—with greater numbers and more destructive technology—subdued the indigenous peoples, stole their land, and systematically destroyed their cultures. But these same invaders, on landing on foreign shores, began immediately destroying the environment through economic exploitation for personal gain, something the aboriginal peoples were not prone to do. In fact, the invaders even fought wars among themselves over who was going to get which of the stolen spoils and how much.

There is a great contradiction here in the notion of *development*. Those invading peoples who deemed themselves more advanced or more developed than the indigenous peoples destroyed lifestyles that had been sustainable, more often than not, for millennia, simultaneously introducing lifestyles of exploitation for personal economic gain that have proven to be nonsustainable. If, therefore, social-environmental sustainability is added as a necessary component to the concept of development in the broader sense, the indigenous Canadians have an even greater claim to being the First Nations, and so do all other indigenous peoples in the world.

But who, then, are the second- and third-world countries? Professor Ralph Metzner, of the California Institute of Integral Studies, had a good idea. He suggested that the world of modern cities and the nation state is the second-world country, while the global, capitalist-industrial economy constitutes the third-world country. Historically, he said, each of these worlds was *superimposed* on earlier cultures (in the sense of absolute force). In an ecological sense, these later, larger systems became parasites that destroyed the indigenous cultures they parasitized. This parasitism was—and is—largely in terms of the flow of energy. "The flow of resources, including raw materials and food," observed Metzner, "is primarily from the indigenous world to the urban, national, and … industrial worlds, whereas military and political control is exerted in the opposite direction."[9]

Sustainable development is thus about the notions of *enoughness* and *repairability*. Here, the operative questions are: When is enough, enough? If we err in our decision, to what extent is the outcome repairable? Such questions are crucial because sustainable development is necessary to promote a change in both the context and content of decisions that affect social-environmental planning and implementation. What is needed to resolve our social-environmental problems goes beyond environmentally safe commodity production and technology.

Instead of the current tinkering with symptoms of our social-environmental malaise, problems need to be solved at their source—worldview assumptions and values—because these drive our decisions, policies, and plans. *Social-environmental sustainability*, on the other hand, questions the very purpose of society and our participation with our home planet and demands social-environmental justice, a challenge going to the very heart of our perceived relationship with Nature and one another, present and future.

We, as planetary citizens, must learn to think at least seven generations ahead when making decisions because the great and only gift we have to give those who follow is *potential choices and some things of value from which to choose*. Today's decisions become tomorrow's consequences, a notion that highlights the word *responsibility*.

Responsibility is a double-edged sword in that our responsibility, our moral obligation, is to choose carefully today so that the generations to come can respond viably to the circumstances we have created for their time of choice. Intelligent decisions on our part are possible only when we both recognize and accept the intrinsic value of Nature as a living organism rather than viewing Nature only as a collective resource (host) from whose body we extract (parasitize) a variety of commodities as the life's blood of our current, dysfunctional, linear economic system.

Development has to be flexible and open to community definition because the values promoted must meet various needs and situations in space and time while safeguarding sustainability. The process of valuation embodied in sustainable development addresses social-environmental justice in recognizing the necessity of equal access to resources, including equal distribution of goods and services, while simultaneously protecting the long-term ecological sustainability of the system that produces them.

Sustainable development also addresses the need to promote education and feelings of self-worth in people, allowing them to act as catalysts in the process of change, whether in their own lives or in the life of society. For change to be a creative process, each person is required to respect every other person as well as the intrinsic value of their environment.

Finally, the valuation/decision process needs to promote the democratic frame of reference because democracy only works when it is actually practiced. In this sense, most of the change has to be directed by the people from the bottom up—the "grass roots" of the local community in whichever country or nation that community happens to be located.

People both define their local communities and are defined by them in that communities play a primary role in maintaining cultural values within and among generations. The collective of individual values determines familial values; the collective of familial values determines community values, which in turn determine appropriate behavior, poverty, and prosperity. As we grow up and are taught at home, educated in schools, and participate in community, socialization occurs, norms are set, and societal control takes place.

People make most of their decisions, do most of their consuming and waste production, and develop many personal relationships within their local community, so it is not surprising that lifestyle becomes a political issue.

Citizens at the local level can also begin drawing connections between personal consumption and its effects on all levels of economic well-being and environmental health. And, because we, as individuals, collectively comprise local communities that, in the collective, comprise society at the regional, state, and national levels, we can heal our global environment simply by changing our individual behaviors. Local communities, as the force that drives change for better or worse, are thus the appropriate scale for dealing with sustainable development through wise, insightful social-environmental planning and implementation.

Summation

In Chapter 11, I discussed the major shift taking place in the world today, a shift that is beginning to more evenly distribute some of the concentrated social wealth—jobs—from the industrialized West, especially the United States, to the rest of the world. Although the redistribution of jobs is for the wrong reason (the corporate money chase in the form of economic competition), it may well have the long-term beneficial effect of stimulating international social parity. Another effect will be a forced reevaluation of the term *development* as it applies both to peoples and nations around the world *and* to social-environmental planning at the local level. Chapter 12, in turn, is a journey through the basic concepts of social-environmental planning in the true sense of sustainable development.

Notes

1. George Bernard Shaw. http://www.goodreads.com/quotes/show/85343 (accessed January 24, 2009).
2. The preceding two paragraphs on outsourcing are based on Paul F. deLespinasse. Flow of U.S. jobs to foreign basses does have its benefits. *Corvallis Gazette-Times* (December 11, 2003).
3. Corvallis Gazette-Times. Greenspan warns against some job-loss cures. *Corvallis Gazette-Times* (February 21, 2004).
4. Ralph Waldo Emerson. http://www.goodreads.com/author/quotes/12080.Ralph_Waldo_Emerson?page=8 (accessed January 26, 2009).
5. Brother David Steindl-Rast. *Gratefulness and the Heart of Prayer: An Approach to Life in Fullness*. Paulist Press, Ramsey, NJ (1984).
6. Victoria Moran. *Shelter for the Spirit: How to Make Your Home a Haven in a Hectic World*. Harper Collins, New York (1997).
7. Brother David Steindl-Rast. *Gratefulness*.
8. Henry David Thoreau. http://quotationsbook.com/quote/42321/ (accessed January 24, 2009).
9. Ralph Metzner. Where is the first world? *Resurgence*, 172 (1995):126–129.

12

Social-Environmental Planning in Space and Time

> To accomplish great things, we must not only act, but also dream; not only plan, but also believe.
>
> **Anatole France**[1]

Our Daily Relationships

There are four basic relationships that underpin social-environmental planning: (1) intrapersonal, (2) interpersonal, (3) between people and the environment, and (4) between people in the present and those of the future. Because everything we do in life is a practice of relationships, it is imperative to social-environmental planning that we both understand these basic relationships *and* strive to become psychologically mature adults within our respective spheres of influence.

Intrapersonal

An intrapersonal relationship exists *within* a person. It is an individual's sense of spirituality, self-worth, and personal growth. In short, it is what makes that person conscious of and accountable for their own behavior and its consequences. The more spiritually conscious a person is, the more other centered the person is, the more self-controlled personal behavior is, and the greater is the person's willingness to be personally accountable for the outcome of their behavior with respect to the welfare of fellow citizens and Earth as a whole—present and future.

Suppose, for instance, someone in a store is rushing blindly to get somewhere and shoves you out of the way; you have a choice of how you respond to being shoved. You can get angry and impatient and say something nasty, or you can be patient, kind, and understanding. Your thoughts and actions are the seeds you sow each time you make a choice of how you treat another person, your interpersonal relationship.

Interpersonal

With respect to our interpersonal relationships, we always have a choice—and we must choose. If we do not like the outcome of our choice, we can choose to choose again. We are not, after all, victims of our circumstances but rather the products of our choices. The more we are able to choose love and peace over fear and violence, the more we gain in wisdom and the more we live in harmony, both within ourselves and with Nature. This is true because what we choose to think about determines how we choose to act, and our thoughts and actions set up self-reinforcing feedback loops that can be thought of as self-fulfilling prophecies that become our individual and collective realities.

It is just such self-reinforcing behavioral feedback loops—based on competition for money through the exploitation of resources—that have made our shared environment a battlefield. Our overemphasis on competition in nearly everything fosters the material insecurity that often manifests as greed.

Another tendency of human beings faced with a perceived threat to their sense of material survival is to defend a point of view. Nonetheless, there are as many points of view as there are people, and everyone is indeed right from their vantage point. Accordingly, no resolution is possible when people, individually or collectively, are committed only to winning agreement with their position. The alternative is to recognize that "right" versus "wrong" is a personal judgment about human values and is not a winnable argument.

For us in the United States at least, a crisis is too often in our point of view because we tend to perceive the world through a disaster mentality, regardless of evidence to the contrary. We are inclined to focus on and cling to a view of impending doom, in part because of our emotional discomfort with an unknown future constantly heightened by daily news with its graphic portrayals of fraud, violence, and tragedies worldwide.

Fear (including "greed" and most other negative personal attributes) is a projection into a future of unwanted possibilities that breeds weakness—a state that leaves little time or energy to develop other areas of a person's life. Out of the irrational logic of fear, people too easily and too often resort to violence toward one another and Earth in an effort to assert the little power they feel is still theirs.

The instantaneousness of today's news does not give us time to assimilate the stories within the context of global proportions and so augments our growing sense of helplessness. News, which came more slowly when I was a boy, could be kept in better proportion relative to its time and area of coverage. Today, on the other hand, newsworthy disasters all seem to happen instantly in our homes via television and can become so overwhelming that we are emotionally drained and numbed by them even as they compound our fears of our own unknowable future. In addition, insurance companies continually foster the disaster mentality.

In a sense, insurance companies are the casinos betting, based on carefully calculated *probabilities*, that nothing will happen to you as they take your money, and that you, by purchasing insurance, are the gambler, betting blindly on the *possibility* that a disaster will befall you at any time. You therefore are betting against yourself—every moment of every day. And, it is just this disaster mentality that causes many frightened people to become increasingly self-centered, wanting everything—now. Unyielding self-centeredness represents a narrowness of thinking that prevents cooperation, coordination, positive possibility thinking, and the resolution of issues.

To redesign and repair our social-environmental relationships and their feedback loops, however, requires teamwork. Setting aside egos and accepting points of view as negotiable differences while striving for the common good over the long term are necessary for teamwork. Teamwork demands the utmost personal discipline of a true democracy and is the common denominator for lasting success in any social endeavor.

But, even if we exercise personal discipline in dealing with current social-environmental problems, most of us have become so far removed from the land sustaining us that we no longer appreciate it as the systemic embodiment of continual processes. Instead, we are symptomatic in our worldview in that we focus on a chosen product by which to measure the success of our *management* efforts, and anything diverted to a different product is considered a failure. It is time to reevaluate our philosophical notions of Nature, community, and society and how they can be sustainably integrated into a common vision for the future.

Between People and the Environment

Sustainability means that, to the extent possible, development programs integrate the local people's requirements, desires, motivations, and identity in relation to the surrounding landscape. It also means that local people, those responsible for development initiatives and their effect on the immediate environment and the surrounding landscape, not only be allowed but also required to participate equally and fully in all debates and discussions, from the local level to the national. Here, a basic principle is that programs have to be founded on local requirements and cultural values *in balance* with those of the broader outside world.

Some years ago, I attended a meeting on the development of rural communities during which economic diversification was the sole focus of discussion. It soon became apparent that the group had no idea of the importance of landscape to the identity of a community. By way of example, a logging community is set within a context of forest, a ranching community is set within the context of lands for grazing livestock, and a community of commercial fishers is usually set along a coastline, be it a great lake or an ocean.

The setting of a community helps define it because people select a community for what it has to offer them within the context of its landscape in conjunction with their sense of values. The location therefore helps create many characteristics that are unique to the community. By the same token, the values and development practices of a community alter the characteristics of its surrounding environment.

In addition to the surrounding, natural environment, the constructed environment within a community is also part of its setting and therefore its identity. This includes the design of buildings, zoning, the configuration of transportation systems, and the allowance of natural occurrences within the structured setting.

In turn, a community's worldview defines its collective values, which in turn determine how it treats its surrounding landscape. As the landscape is altered through wise use or through abuse, so are the community's options altered in like measure. A community and its landscape are thus entwined in a mutual self-reinforcing feedback loop as the means by which their processes reinforce themselves and one another.

Each community has physical, cultural, and political qualities that make it unique and more or less flexible. The degree of flexibility of these attributes is important because sustainable systems must be ever flexible, adaptable, and creative. The process of sustainable development must therefore remain flexible since what works in one community may not work in another, may work for a different reason, or may just work differently.

Beyond this, the power of sustainable development comes from the local people as they move forward through a process of growing self-realization, self-definition, and self-determination. Such personal growth opens the community to its own evolution within the context of the people's sense of place, as opposed to coercive pressures applied from the top down or from the outside inward.

Sustainable development encompasses any process that helps people meet their requirements, from self-worth to food on the table, while simultaneously creating a more ecologically and culturally sustainable and just society for the current generation and those that follow. Oddly, perhaps, but a wonderful example of people getting in touch with their self-worth can be found in "ecopsychology" as practiced in the Natural Growth Project in North London, England, and is part of the Medical Foundation for the Care of Victims of Torture.

People who have been badly injured by other humans as well as dislocated from their homes and country often find it easier to approach Nature rather than risk a relationship with other people. As part of their therapy, they are given community allotments in which to dig in the soil, plant seeds, nurture the plants, weed, and compost—all of which provide metaphors for tending the human soul. This kind of therapy is a fundamental tenet of ecopsychology, which is based on the belief that our current social dilemma is a result of the Western Judeo-Christian paradigm in which believers regard themselves as separate from and above Nature. In fact, the Judeo-Christian attempt to subjugate Nature dominates virtually all Western social systems—from religion to education, economics, and politics. In all these social systems, there seems to be a deep-seated fear of the "wild, untamed" world, of being overwhelmed by primordial, uncontrollable instincts.

Although the philosophical foundation of ecopsychology is beyond the scope of this book, one could say that it works toward restoring health rather than suppressing the symptoms of illness. Author Mary-Jayne Rust put it this way:

> Discovering our yearning, in all its nakedness, is the central part of the therapeutic process, for our yearning is the rudder of our lives. It emanates from the opening of our hearts and shows us where to go next. If we dare to follow, and we can endure the path in all its joy and pain, our compassion and wisdom grow. We extend our capacity to identify with the pain of the "other," be they human or not.[2]

In the end, a sign of psychological maturity, which equates to mental/emotional health, is our capacity to recognize and account for what best serves the whole in any given situation. In turn, all social-environmental planning and development require psychological maturity if they are to be truly sustainable. I say this because sustainability has to integrate the requirements of a local community with those of the immediate environment and surrounding landscape while instilling a relative balance between the local community and the larger world of which it is an inseparable part in space and time.

Between People in the Present and Those of the Future

We, and our leaders, are now obliged to address a moral question: Do those living today owe anything to the future? If our answer is "No," then we surely are on course because we are consuming resources and polluting Earth as if there were no tomorrow. But, on the other hand, if the answer is "Yes, we have an obligation to the future," then it is incumbent on us to determine what and how much we owe because our present, nonsustainable course is rapidly destroying the environmental options for all generations. Meeting this obligation will require a renewed commitment to the highest human ideals from all walks of ideology—to do unto those to come as we wish those before us to have done unto us.

To change anything necessitates reaching beyond where we are, beyond where we feel safe. We must dare to move ahead, even if we do not fully understand where we are going or the cost of getting there, because we will never have perfect knowledge. In addition, we have to ask innovative, other-centered, future-oriented questions in order to make the necessary changes for the better.

True progress toward a biophysically sound environment and an equitable world society will be expensive in human effort (discipline) and money with which to heal the environmental damage done by the centuries of unbridled economic competition. But, the longer we wait, the more disastrous becomes the social-environmental condition and the more expensive and difficult become the necessary social changes. For this reason, it behooves us to begin approaching true long-term social-environmental planning now, commencing with the notion of a "living trust."[3]

Everyforest and Everycity as a Living Trust

Although most people speak of "stewardship," the concept of a "living trust" is preferable because stewardship does not, in and of itself, have a legally recognized "beneficiary"—someone who directly benefits from the proceeds of one's decisions, actions, and the outcomes they produce. Although a *steward*, by definition, is someone who "manages" another's property or financial affairs and thereby acts as an agent in the other's stead, there is nothing explicit in the definition about a legal beneficiary. For this reason, stewardship seems the weaker of the two terms because the fiduciary responsibility of stewardship is to the shareholders, whereas the fiduciary responsibility of a living trust is to the beneficiaries, none of whom need to be physical shareholders.

A living trust is like a promise, which in our case is to keep the social-environmental feedback loops in good repair. "Promises are scary things," said author Elizabeth Sherrill. "To keep them means relinquishing some of our freedom; to break them means losing some of our integrity. Though we have to make them *today*, promises are all about *tomorrow*—and the only thing we know for sure about tomorrow is that we don't know anything for sure!"[4]

A living trust, in the legal sense, is a present transfer of property, including legal title, into trust, whether real property (such as forestland or a historical building) or personal property (such as interest in a business). The person who creates the trust (say, the owner of forestland or a historical building) can watch it in operation, determine whether it fully satisfies their expectations, and if not, revoke or amend it.

A living trust also allows for the delegation of administering the trust to a professional "trustee," something that is desirable for those who wish to divest themselves of managerial responsibilities. The person or persons who ultimately receive the yield of the trust, for better or worse, are the legal beneficiaries. The viability of the living trust is the legacy passed from one generation to the next.

Although a trustee may receive management expenses from the trust, meaning that a trustee may take what is necessary from the interest, at times even a small stipend, the basic income from the trust as well as the principle must be used for the good of the beneficiaries. Yet, natural resources in our capitalist system are *assumed* to be income or revenue rather than capital. That said, a trustee is obligated to seek ways and means to enhance the capital of the trust—not diminish it. Like an apple tree, one can enjoy the fruit thereof but not destroy the tree. A living trust, after all, is about the quality of life offered to the generations of the future; it is *not* about the acquisition of possessions. Thus, "real learning" is a requisite of caretaking a living trust.

Real learning—the remembrance of things forgotten and the development of things new—occurs in a continuous cycle. Learning encompasses theoretical and practical conceptualization, action, and reflection, including equally the realms of intellect, intuition, and imagination. Real learning is important because overemphasis on action, one part of which is competition, simply reinforces our fixation on short-term, quantifiable results. Our overemphasis on action precludes the required discipline of reflection, a persistent practice of deeper learning that often produces measurable consequences over long periods of time.

Many of today's problems resulted from yesterday's solutions, and many of today's solutions are destined to become tomorrow's problems. This simply means that our quick fix, social trance blinds us because we insist on little symptomatic ideas that promote fast results, regardless of what happens to the system itself. What society really needs are "big fixes" in the form of systemic ideas that promote and safeguard the long-term integrity of our social-environmental relationships (e.g., a collective vision of Everyforest as a biological living trust and Everycity as a cultural living trust).

A Living Trust as a "Big Idea"

Where, asked the late publisher Robert Rodale, are the big ideas, those that change the world? They probably lie unrecognized in everyday life since our culture lacks sufficient free spaces for unfettered thought. A "big idea," according to Bob Rodale, has seven basic characteristics. Let us see if the idea of a living trust satisfies them:

1. It (the idea) has to be generally useful in good ways. A living trust translates into a healthy environment and available resources, both natural and cultural.
2. It must appeal to generalists and give them a leadership advantage over specialists. A living trust requires an understanding of the whole system *and* so necessitates an amalgamation of generalists and specialists, with generalists in charge.
3. It has got to exist in both an abstract and a practical sense. A living trust, as seen in item 1, is practical in its outcome, but it is also abstract in that its practical outcome requires people to work together with respect, humility, wonder, and intuition as well as their varied intellectual gifts.
4. It requires some interest at all levels of human concern. A living trust involves the continual building of relationships, which is all we humans really do in life, and so touches all levels of society, both within itself and with Nature.

5. It has to be geographically and culturally viable over extensive areas. A living trust is essential if the natural and social world is to remain viable and habitable for the generations of the future.
6. It necessitates encompassing a multitude of academic disciplines. To caretake a living trust requires the integration of all disciplines, such as soil science, mycology, philosophy, sociology, theology, education, politics, ecology, forestry, architecture, urban planning, and economics.
7. It entails having a life over an extended period of time. A living trust is, by definition, an instrument of continuity among generations.[5]

A living trust seems to fit all of Rodale's requirements in that the quality and health of our social-environmental partnership is everyone's birthright and so the foundation of the global commons. The collective criteria of a living trust as a big idea also helps people to understand that life is not condensable, that any model is an operational simplification—a working hypothesis that is always ready for and in need of improvement. When we accept that neither shortcuts nor concrete facts exist—in that knowledge is merely some version of the truth—we will see how communication functions as a connective tool through which we can and must share experience, invention, cooperation, and coordination.

When people speak from and listen with their hearts, they unite and produce tremendous power to invent new realities and bring them into being through collective actions. While today's environmental users with narrow special interests will not be around in forest and city by the end of this century, all of the environmental and cultural necessities will be, and that makes trusteeship critically important.

Trusteeship is a process of building the capacity of people to work collectively in addressing the common interests of all generations within the context of sustainability—biologically, culturally, and economically. A biological living trust in turn means honoring the productive capacity of an ecosystem within the limitations of its biophysical principles.[6] A cultural living trust means honoring the fact that each building has, from the first cave inhabited by a hominid, been a functional text in humanity's architectural library of time. Each structure is thus a volume in the library, and language is the librarian that guards the knowledge harvested and stored through the count of ages.

This said, if we begin now to caretake forests as a biological living trust and cities as a cultural living trust, that *is* a big idea. After all, social-environmental sustainability is only a choice—our choice—but one that demands careful and humble planning if it is to endure the often shortsighted, contradicting political vagaries of humanity.

A living trust (whether biological or cultural) is predicated on systemic "holism" in which reality consists of an organic and unified whole that is greater than the simple sum of its parts. That is, the desired function of a system defines its necessary structure. The structure in turn defines the required composition that creates the structure that allows the functional processes to continue along their designated courses. Consequently, wisdom dictates that we learn to characterize a system by its function, *not its parts.* The basic assumptions underpinning a living trust (whether biological or cultural)—all externalities within the current economic framework of forestry and city planning—are:

- Everything, including humans and nonhumans, is an interactive, interdependent part of a systemic whole.
- Although parts within a living system differ in structure, their functions within the system are complementary and benefit the system as a whole.
- The whole is greater than the sum of its parts because how a system functions is a measure of its biophysical integrity and biological/cultural sustainability in space through time.
- The biophysical integrity and biological/cultural sustainability of the system are the necessary measures of its economic health and stability.
- The biophysical integrity of processes has primacy over the economic valuation of components.
- The integrity of the environment and its biophysical processes have primacy over human desires when such desires would destroy the system's integrity (productivity/cultural history, e.g., identity) for future generations.
- Nature determines the necessary limitations of human endeavors.
- New concepts need to be tailored specifically to meet current challenges because old problems cannot be solved in today's world with old thinking.
- The disenfranchised, as well as future generations, have rights that must be accounted for in present decisions, actions, and potential outcomes.
- Nonmonetary relationships have value.

In a biological/cultural living trust, the behavior of a system depends on how individual parts interact as functional components of the whole, not on what each part, perceived in isolation, is doing. The whole in turn can be understood only *through* the relationship/interaction of its parts. Hence, to understand a system, as stated in Chapter 3, we need to understand how it fits into the larger system whereof it is a part. This understanding gives us a view of systems supporting systems supporting systems, ad infinitum. Consequently, we move from the primacy of the parts to the primacy of the whole, from insistence on absolute knowledge as reality to relatively coherent interpretations of constantly changing knowledge, and from an isolated personal self to self in community.

Remember, to protect the best of what we have in the present for the present *and* the future requires us to continually change our thinking and our behavior to some extent. Society's saving grace is that we all have a choice. Accordingly, whatever needs to be done can be done—if enough people want it to be done and decide to do it.

Everyforest Is a Biological Living Trust

Because a forest is a living entity that requires nothing from humanity to fulfill its intrinsic purpose, it can be thought of as a "*perpetual*, biological living trust" (hereinafter referred to as "biological living trust") in which individual people—as well as their relationships among one another, Nature, their communities, and generations—have value and are valued, as are all living beings. In this sense, a forest is the background, the context of the social-environmental feedback loops that supports and sustains the foreground represented by a city.

For forestry to survive the twenty-first century as an honorable profession, therefore, the moral essence of a biological living trust needs to be accepted. The profession must also advance beyond resisting change as a condition to be avoided (clinging to the current linear, reductionistic, mechanical worldview of exploitive resource extraction) and embrace change as a process filled with hidden, viable biophysical-social-economic opportunities in the present for the present *and* the future—the beneficiaries.

If we have the courage and the willingness to adopt and implement the concept of a biological living trust, we are practicing sustainable forestry wherein ever-adjusting relationships—biophysical, social, and economic—become the creative energy that guides a vibrant, adaptable, ever-renewing forestry profession through the present toward the future. After all, forestry could be a profession that constantly opens the mind with growing conscious awareness because the forests of tomorrow will be created out of the inspirations, discernment, choices, decisions, and activities of today. In addition, sustainable forestry honors the integrity of both society (intellectually, spiritually, and materially) and its environment, thereby fitting the concept of a biological living trust in that it *maintains* positive outcomes for both the forest as a dynamic system and the beneficiaries who depend on the forest to nurture their well-being in the city.[7]

Everycity Is a Cultural Living Trust

From its dawning, humanity has sought the roots of meaning, the purpose of existence. Having found a bit of meaning here and a tad of purpose there, those who went before us have passed their collective understanding to us in the form of stories, rituals, myths, customs, and the social values stored in each—values pieced together over time to reflect the meaning and purpose of life, the essence of being human. Today, we term the integration of these bits and pieces *culture*.

Whereas most forests come and go, as a few mighty civilizations and their great cities have done, other cities have outlived the original forest that once nurtured them. In the former case, both forest and city have faded from visible participation in the social-environmental tapestry and now reside in the ethers of memory. Other cities remain today in the foreground of our human journey as cultural living trusts supported by and nurtured by their biophysical backgrounds.

Every city is a cultural living trust in that culture is the collective human experience passed forward to inform the present generation of its origins and inspire its future possibilities. Culture is a road map of meaning that leads to a city, where it guides us in the present toward the future along a path that others have already trod; thus, we know we are not alone. Moreover, culture gives us a sense of continuity and thereby helps to ensure that our life, as we interpret it, has added at least a modicum of meaning to the human journey. In a case such as this, culture is aided by the existence of trees.

Many older residents of the U.S. Virgin Islands know there are spirits living in the baobab tree. Its massive, columnar trunk reaches toward the sky, where its branches interlace like fingers. "When all the world was made, the baobab was the last tree created." That is what the grandmothers say. "Jumbies, the undead, love to hide there."

Because a large baobab is very old and comprised mostly of water, when it dies, the water evaporates; its wooden body turns into dust and blows away. Like human flesh, the old tree comes from the mystery of soil and returns to the mystery of the soil. That is but one of the reasons spirits live there.

The Virgin Islanders have known this truth for centuries, but how long will they remember it? Will the younger generations even care? What will happen to these stories when all of the grandmothers and the old trees are gone?[8]

Although there may not be such vivid stories about the old trees in our cities, they are, nevertheless, living beings that have seen the past and have carried it forward with them into the future, to a time we call the present. Here, in this moment, we can share the presence of these old trees and thus muse about the foretime, but only if we know where they are and only if they are protected from thoughtless destruction in the name of economic development.[9]

Many an old tree in a city was there prior to a building or a bridge and has outlasted them. Yet together, the old trees and old buildings of today are the sentinels, the historians of another time when life was slower and, perhaps, more peaceful. In like measure, what we plant and build today will, in some distant time, be the sentinels, the historians of our time carried through the decades and centuries into an unknown and unknowable future.

What will the people in that distant present think of us and what we did with the time we had allotted? Will they think we administered their cultural living trust wisely? I hope so because administering wisely the cultural living trust that now rests in our care is one of the sacred purposes of life—our gift to the generations yet unborn, a gift of simultaneous understanding of who *we were* and who *they are*. To administer wisely the cultural living trust in our care requires long-term social-environmental planning.

Long-Term Social-Environmental Planning

Benjamin Franklin was of the opinion that doing 1,001 small things right would add up to doing a big thing right. Whereas that notion is an excellent motto for long-term social-environmental planning, it must include understanding long-term biophysical trends.

A *trend* is a line—albeit not a straight line—of general direction or movement in the environment defined by a multitude of interacting factors, including the following:

- location of event (e.g., on land, in water, in air, in the tropics, at the North or South Pole, in a valley)
- size of event (e.g., in someone's backyard, on an acre [hectare], over a landscape, over a continent)
- duration of event (e.g., ten seconds, an hour, a year, a century, a geological epoch)
- time of event (e.g., day, night, season, year)
- frequency of occurrence (e.g., once, hourly, daily, seasonally, annually)
- distance between events, such as an inch (2.5 centimeters), a foot (30.5 centimeters), a yard (0.9 meters), a mile (1.6 kilometers), 1,000 miles (1,610 kilometers), 10,000 miles (16,000 kilometers)
- uniformity of event (e.g., uniform, roughly connected, disjunct)
- type of event (e.g., physical, biological, political, a combination)

The infinite variety of interactions among these factors creates an infinite variety of short-term trends that fit into a longer-term trend that fits into a still longer-term trend ad infinitum. Studying short-term trends (those that can be detected and perhaps understood) and projecting them over time may allow some degree of predictability of Nature's responses to our decisions and actions.

There are two cautions, however. First, we must accept that all of these trends are ultimately cyclic, and that their governing principles are neutral and impartial; so, the shorter the trend the more imperative is our acceptance of Nature's neutrality and impartiality. Governing principles, whether biological or physical, are always neutral and impartial.

On the other hand, when we, through politics, assign values to Nature's actions based on our perceptions of "good" and "bad," we interject the artificial variable of partiality that, all too often, clouds our vision and so our judgment. When this happens, we rob ourselves of our ability to predict what might happen in the future with any degree of accuracy.

For example, a hillside meadow is seen as having no value if it bakes under the summer sun and turns brown because it does not produce a visible product the community wants. Accordingly, the meadow is subdivided for a housing development. Within a year of completing the housing development, many of the community's wells begin drying up because the meadow had in fact been the water catchment that unobtrusively supplied the wells.

Second, short-term trends have to be viewed in relation to long-term trends and long-term trends in relation to even longer-term trends. The more we trace the present into the past, the better we understand the present. The more we project the present into the future, the more humble we need to be in our certainty that we understand the present. Knowledge of the past tells us what the present is built on and on what the future may be projected. But, this is true only if we accept past and present as a cumulative collection of our understanding of a few finite points along an infinite continuum—the trend of the future, which is an unmitigated abstraction to many people.

Thinking and knowledge in Western society have become so linear we have forgotten that everything is defined by its relationship to everything else. Nothing exists by itself; everything exists in relation to something. In the example above, the community's well water existed in relationship to the hillside meadow and the precipitation that fell on it. But, the relationships were not understood until the housing development irreparably altered them—and perhaps not even then.

Failing to account for a community's long-term supply of water in the face of short-term dollars to be made by a few people is dangerous because changes in the spatial patterns of land use, which grossly alter habitats through time, may well be crucial to understanding the dynamics of landscapes and will have implications for many biophysical processes. Changes in the patterns of landscapes are related to the flows of materials and energy across them, such as the processes of erosion and the movement of water and sediments. Characterization of the relationships among changing patterns and how those changes affect biophysical processes is particularly important if we are to develop a more complete understanding of landscape dynamics, our effects on them, and their effects on our communities.

Here, it may be wise to take a lesson from the U.S. military, albeit the military operates with a perpetual disaster mentality. Nevertheless, military planners use their disaster mentality to examine possible worst-case, "what if" scenarios in order to evaluate the military's state of preparedness.

We need to do the same what if scenarios in long-term social-environmental planning in a peaceful society, but with a positive outlook, making it necessary to revisit Benjamin Franklin's notion of doing 1,001 small things right in order to do a big thing right. And, that "big thing" will be our legacy to all generations, although we will not live to see the outcome of our long-range planning.

Planning Increments

In regard to "planning increments," Bruce E. Tonn, a research scientist at the Oak Ridge National Laboratory in Oak Ridge, Tennessee, referred to long-range planning as "500-year planning."[10] While 500 years will likely be an abstraction to most people, it can be approached by viewing the 500-year timescale in five increments: (1) five years, (2) ten years, (3) twenty-five years, (4) one hundred years, and (5) five hundred years.

Five years. Why five years? A five-year increment is easier to deal with in terms of decades and centuries than would be the seven-year increment business sometimes uses.

Ten years. This decadal measure is perhaps the most meaningful to the average person because we tend to reflect on our lives each time we reach a decadal birthday. In addition, a decadal increment both deals with the easily computed numerical measure of ten and simultaneously reflects the projected outcome to two financial planning cycles. Since economics is the language of industrialization, it is an excellent medium through which to translate scientific principles to the nonscientific public in a way that allows them to learn while retaining their dignity intact.

Twenty-five years. Twenty-five years represents a generation, the average period between the birth of the parents and the birth of their offspring. The generational increment is important from an educational point of view, something I learned in Japan while working with the Shinto priests.

About thirteen hundred years ago, Emperor Tenmu ordained the practice of removing an old Shinto shrine and rebuilding a new, exact replica next to it every twenty years. It is not clearly known why Emperor Tenmu stipulated the rebuilding of the shrine at this interval, but it is likely that twenty years was thought to be the optimum period for two reasons. First, it was to preserve the Grand Shrine because it has a thatched roof, unpainted or otherwise preserved structures, and is erected on posts sunk into the ground with the benefit of foundation stones.

Second, twenty years is perhaps also the most logical interval in terms of passing from one generation to the next the technological expertise needed for the exacting task of duplicating the shrine because a man of twenty works as an apprentice when he helps to rebuild the shrine for the first time. At age forty, he is already versed in the art of replacing the shrine but still has more to learn. By age sixty, he is the master carpenter who passes the legacy to the twenty-year-old and forty-year-old apprentices working at his side. This trigenerational method of replacing the shrine every twenty years is critical because the architecture represents the symbolic librarian of the Shinto belief—the cultural knowledge that has been passed from one generation to the next for thirteen hundred years without change, without written text, and will continue so into the future.

One hundred years. A century is roughly three generations as well as a time frame long enough to witness the emergence of some biophysical thresholds, both good and bad in terms of humanity, while still having time to make necessary behavioral corrections toward the five-hundred-year goal of social-environmental sustainability.

Five hundred years. Five hundred years is an excellent goalpost because Nature, for the most part, moves far more slowly than we are accustomed. As our pace of life becomes faster, we become more impatient and want things to go still faster. As we become more organized, we become less spontaneous and hence less joyful. We are better prepared to react to some aspects of the future but less able to enjoy all aspects of the present or to *reflect* on the past. And, wisdom requires reflection so we can *respond* rationally and wisely to unwanted circumstances, as opposed to *reacting* with irrational logic.

"As the tempo of modern life has continued to accelerate," said author Jeremy Rifkin, "we have come to feel increasingly out of touch with the biological [and spiritual] rhythms of the planet, unable to experience a close connection with the natural environment."[11] Our human perception of time is no longer joined to the flow and ebb of the tides, the rising and setting of the sun, or the eternal parade of the seasons. We have instead created an environment governed by artificial time punctuated by electronic impulses from the heart of technology. But technology, contrary to the thinking of many people, is not the culprit in our ever-faster pace of life; the culprit is economics.

To Wolfgang Sachs, a prominent German environmental thinker, speed is an unrecognized factor that fuels environmental problems. It is possible, contended Sachs, to talk about the ecological crisis as a collision between scales of time—the fast scale of human modernity crashing into the slow scale of Nature and Earth.[12] A Chinese proverb puts it this way: The oxen are slow, and the Earth is patient. In our fast-paced world, Sachs said, we put more energy into arrivals and departures than we do into the experience itself. In this sense, our obsession with constant motion virtually guarantees that we miss the very experience we are rushing to meet.

The speeding internal clock of our constantly go-faster society is difficult to escape even as it precludes most of us from thinking in terms of consciously varying the pace of our lives to find therein the hidden beauty since our culture deems speed to be "productive." It is, nevertheless, imperative to break the sense of time as taskmaster because raising children, making close friends, creating works of art, and leaving a positive social-environmental legacy for our heirs all require various scales of time since all are practices in the aesthetics of relationship.

The aesthetics of relationship in scales of time is nicely illustrated in a story Gregory Bateson told about the replacement of the gigantic oak beams in the ceiling of one of the dining halls at New College, Oxford University, in

England. Some years ago, an entomologist went up to the roof of the dining hall and, poking the huge oak beams with a pocketknife, found them to be full of beetles. So, members of the college council called in the college forester and asked him where they could find oak beams of like quality. "Well, sirs," he said, "we was wonderin' when you'd be askin'." On further inquiry it was discovered that a grove of oaks had been planted to replace the beams in the dining hall when the college was founded them infested with beetles, probably in the late sixteenth century. Planting the oaks had been deemed necessary because the oak beams have a propensity to become "beetly" in the end. Consequently, the plan of saving the oaks for the dining hall had been passed down from one forester to the next for four hundred years. "You don't cut them oaks. Them's for the College hall."[13] By protecting the oaks through the centuries, one could say the people who planted them may have understood and appreciated time in its various scales as well as seeing time as a mystery to be contemplated rather than a foe to be vanquished.

Here, it is important to understand that slowness is not in opposition to speed but rather is the middle path between *fast* and *inert*. Therefore, if we give up always looking beyond the task at hand, we will inevitably find the required time for the task and do it well the first time, thus actually saving time and money somewhere in the future.

In short, we in the United States, a nation with the world's greatest wealth of Nature's biophysical capital and services, are summarily squandering the inheritance of future generations—at home and abroad—because we lack the self-discipline to create and implement social policies based on humility and prudence in our local, regional, and national planning, both financial and land use. Speaking as a research scientist who has for many years worked to uncover and understand the social-environmental relationships that govern the potential quality of human life, I know that a goodly number of the fundamental biophysical feedback loops are known and understood and can be accommodated in all types of social-environmental planning if we so choose. It is, after all, only a choice, but a choice that demands the discipline to act in the spirit of that choice—and thus is *not* a question of can we or can't we, but rather of *will we or won't we*. Moreover, virtually all of our social-environmental problems stem from the current philosophy and practice of market economics, which is the reason the first increment of five-hundred-year goalpost planning is that of the economic timeline. What we in the United States, indeed in the world, need is competent, farsighted leadership to complement our scientific knowledge with the foresight of sound, long-term social-environmental planning—*not* economic globalization. Now, the question is, where to begin?

Where to Begin?

As previously stated, the farther we predict into the trackless future, the more conscious and clear we need to be of our vision, goals, objectives, and data—that to achieve the ideal of our dreams necessitates making one small correction at a time as early in the game as possible. Understanding this lesson is imperative because long-term social-environmental problems are self-reinforcing feedback loops that take a long time to develop in the invisible present before their threshold of effects becomes manifest. Likewise, these negative environmental effects will take a long time to repair (should people even know how) because the particular feedback loop must first be arrested and then altered before any positive shift can take place. Even then, a social-environmental problem that took a century or more to develop will take at least that long to realign, although a complete reversal will be impossible because the biophysical system will have shifted its evolutionary direction in the interim.

So, when and where do we begin? We begin right now, here, at home—wherever home is—by doing four things:

1. Get the leaders to sit down together and reconcile human values and activities with the biophysical realities of Nature's systemic constraints as they affect the ecological services people require.
2. Commence deciphering the ecological patterns in space and time as they affect a community's sustainability *and* the reciprocal image of how the community affects the sustainability of the ecological patterns.
3. *Respond* (as opposed to react) to whatever vexing circumstances arise, something that can be accomplished through transformative conflict resolution.
4. Based on the other three things, conduct an inclusive participatory process to determine the vision, goals, and objectives for the community's long-term social-environmental planning.

Vision

A *vision* is the philosophical vanishing point on a distant horizon; it is the ideal that gives value to life's often-arduous journey. The ideal for long-term planning, as I envision it, is to pass forward a healthy, vibrant Earth that provides humanity with the best opportunity to achieve its highest potential in creating and maintaining social-environmental harmony.

Goals

A *goal* constitutes a clarification of the vision by elucidating a major area of accomplishment that needs to occur if the philosophical underpinning of the vision is to be upheld and realized. In this sense, goals represent the major dimensions of the vision that must be monitored to determine the state of our environment's condition at any given time. It is vital that goals in long-term planning include:

- The arrest and repair of environmental degradation by making small, timely corrections dictated by the best scientific data, thereby minimizing potential human-caused, catastrophic events. Such environmental protection and repair is to be accomplished by modifying long-term human behavior through education and legal mandates because the arrest of social-environmental problems is impossible by quick technological fixes without substantive modification in personal and social behavior.
- The transformation of the current, exploitive socioeconomic system into one that is friendly, or at least benign, to both people and the environment. Such transformation is to be controlled by law or central government planning and direction. Any socioeconomic system that is developed must *debit* as well as credit in order to measure *and* illuminate the "real" cost of doing business.
- The widely accepted understanding that women must have parity with men if the world's human population is to be controlled with any semblance of dignity and longevity. This endeavor will require a long and arduous educational process, one demanding humility on the part of the male population.
- The consciousness of and accountability for the appropriateness of the technology that is created and introduced into the environment, meaning extensive risk analyses based first and foremost on compliance with Nature's biophysical constraints—and last on economic cost-benefit analyses. When such cost-benefit analyses are done, they must include *real costs* to future generations as determined by a system of genuine economic indicators.

Objectives

A vision and its attendant goals describe your desired future condition. As such, they are qualitative and thus not designed to be quantified. An *objective*, on the other hand, is quantitative and so is specifically designed to be quantifiable. To ensure the effectiveness of an objective requires asking: Is the objective specific enough? Are the results clearly quantifiable and within specified scales of time?

Challenges of Long-Term Planning

Long-term planning probably began in a relative sense when the nomadic way of life in hunter-gatherer societies gave way to nomadic herders and then to settlement. Such planning was based around the annual supply of food to gain some control over the vagaries of the environment.

Today, planning is more complicated and more sophisticated than ever but still overly shortsighted when it comes to leadership. Nevertheless, we need *real* long-term social-environmental planning with a five-hundred-year horizon. Although long-term planning, by its very nature, requires a sound philosophical foundation to support its tenets and guide its methodologies, there are some challenges: philosophical, scientific, conceptual, policy, and leadership.

Philosophical Underpinnings

Perhaps the major philosophical difficulty lies in determining what is meant by intergenerational equity. That is to ask, what, if anything, do we in the present adult generations (young and old) owe to those just entering the world and those yet unborn? By its very nature, the concept of intergenerational equity or environmental justice, from the human point of view, asserts that we owe something to every other person sharing the planet with us, both those present *and those yet unborn*. But, what exactly do we have to give?

Here, one needs to understand and accept that all we have to give one another of real value—*ever*—is our love, our trust, our respect, and the benefit of our experience. All we have to give our children and grandchildren as a personal gift is our love, trust, respect, and the benefit of our experience. There is a caveat to this statement, however, because we give our personal gift in two ways: We give individually to a child as guidance for today, and we give collectively with our neighbors as social-environmental options passed forward for the benefit of our children's and grandchildren's tomorrow.

With every option we protect and hold open, we pass forward a choice and something of value from which to choose. Our love, trust, respect, and the benefit of our experience are encompassed in every choice we pass forward. Likewise, these four things are withheld in each option we foreclose. In the end, all we have to give children are choices and some things of value from which to choose. Our gift must be unconditional to meet the values of tomorrow, *not* a judgment based on our values of today.

"Every age and generation," wrote Thomas Paine, "must be as free to act for itself in all cases as the ages and generations which preceded it. The vanity and presumption of governing beyond the grave is the most ridiculous and insolent of all tyrannies."[14] In a similar tone, Thomas Jefferson concluded in 1789 that "the earth belongs in usufruct [meaning trust] to the living, that the dead have neither powers nor rights over it. ... No society can make a perpetual constitution, or even a perpetual law. The earth belongs always to the living generation; they may manage it then, and what proceeds from it, as they please, during their usufruct."[15]

Although Paine's and Jefferson's comments may sound like each living generation has the right to use the world with total abandon and without a thought for future generations, such is not true. Paine's first statement ("Every age and generation must be as free to act ... as the ages and generations which preceded it") says, in effect, that each generation must save and pass forward all possible options to the next generation or the latter will *not* be as free to choose as the preceding one.

Paine's second statement ("The vanity and presumption of governing beyond the grave is the most ridiculous and insolent of all tyrannies") points out that we cannot force on any future generation our current set of values and that to try is both futile and arrogant. Beyond that, his statement berates the corporate mentality that all too often says unless there is a guarantee that the next generation will follow our current set of values, there is no point in saving resources for it. Ergo, those with the corporate mentality quickly exercise the options available to the present generation and liquidate coveted resources *before* the next generation has a voice.

Jefferson's second statement ("No society can make a perpetual constitution, or even a perpetual law") is similar to Paine's second. On the other hand, Jefferson's first statement ("that the earth belongs in usufruct [meaning trust] to the living, that the dead have neither powers nor rights over it), and his third statement ("The earth belongs always to the living generation; they may manage it then, and what proceeds from it, as they please, during their usufruct") clearly indicate that viable governance is obligated to track with the times, meaning people must change as the times do, something we are loath to accept.

Change comes first to local communities and their environments through a sound, equitable, and creative public school system that gives each student an equal opportunity to understand the democratic principles of self-governance as well as the knowledge and values whereon such governance is founded. The very economic, social, and political future of our country and its democracy rests squarely on the shoulders of education and long-term social-environmental planning. With this in mind, Lester Thurow, a well-known economist, stated many years ago that he did not mind paying property taxes because a goodly part is used to finance local school systems, so he was, in essence, protecting his own future.

Our children *are* our future. What we do for them now by providing them with a solid, intelligent, far-sighted, and well-rounded education that stresses social-environmental harmony through long-term planning will rebound many times over to the collective benefit of our present generation and all generations to come. So, why are people reluctant to act?

There are several reasons people refuse to act, not the least of which is the perceived personal risk involved in the early stages of a philosophical revolution, when the risk seems greater than the potential material gain. The foregoing sentence is just another way of saying that people who start a revolution, even a successful one, are seldom the beneficiaries of what they accomplish. Nevertheless, we can choose to live our one unrepeatable life in a bold and spirited way that changes us even as we change the world for the better, albeit minutely in the scheme of things, *or* we can try to cheat death through rampant materialism. Regardless of what path we choose, the outcome is the same—but the event of really living versus merely existing in fear of death's shadow is our choice. And, we *must choose* because there is no such thing as *no choice*.[16]

Scientific Underpinnings

The role of science is to understand the outworking of Nature's biophysical relationships. Even so, science cannot, and will not, provide definitive answers to any of society's questions and cannot be used to derive perfect knowledge with respect to any discipline. But, science can provide a well-grounded understanding of Nature's biophysical causes, effects, and the processes thereof and so greatly increase the probability that we *will know enough* to make wise decisions in the present for the present *and* the future. There is a condition, however; we must accept what science has to offer in the spirit of advancing the knowledge of potential social-environmental outcomes, both positive and negative, rather than defending special interest points of view.

Conceptual Framework

A thought expressed by a Chinese Mandarin to a Chinese army major in the 1958 movie *Inn of the Sixth Happiness* is important to understand in the planning process, despite the fact that on the surface it seems contrary to planning: "A planned life, my friend, is a closed life [in other words, rigid]. It can be tolerated, perhaps, but it cannot be lived."

As members of Western society, we are encouraged to plan as a response to the reductionist model of a world characterized by materialism—the predetermined, piecemeal design. This mechanistic worldview says in effect that if we can control all the pieces of our lives, we can plan for and thus control the outcome of our experiences, an idea that defines us as a cog within a huge machine.

As such, our highest goals are to plot our lives and circumstances to produce the least friction, to create a new and better operating system or a richer, more rewarding feedback loop. These ideas are dualistic, nonetheless, and serve to separate us from other people and from Nature.

This sense of separation creates an image of us as isolated individuals on our home planet, each trying to control events and circumstances in our moment-to-moment experiences, producing an underlying feeling that life is a solitary battle we are losing. We seek security and our sense of identity in our individual accomplishments rather than trusting an interconnected, interactive consciousness.

Lacking a secure sense of identity, we replace spontaneous creativity with a checklist of things to do, thus saying that we cannot trust anyone else to look out for us, that we are forced to look out for ourselves, which leaves us feeling alone and isolated in the universe, often with a palpable feeling of powerlessness. In addressing this lack of trust, Mahatma Gandhi once asked: "Does not the history of the world show that there would have been no romance in life if there had been no risks?"[17] And yet, the sense of risk is what we are constantly trying to minimize, or even eliminate—the sense of adventure, as it were.

"Intention" is a focused purpose, the adventure, that has within it the seed of fulfillment because of that focus. In this sense, intention is different from planning because it expresses the desire and then allows the spontaneous, intuitive outworking of the results. Planning, on the other hand, is the linear attachment to the detail and outcome that often contradicts trust in the spontaneous and the intuitive, thus limiting possible outcomes.

If we do not trust our spontaneous creativity and our intuition—if we do not trust one another to act appropriately in an other-centered fashion—then we feel forced to "plan" an alternative, a contingency. But, Nature offers another perspective.

In the aforementioned Australian savannah are twenty-foot-tall termite towers, offering a magnificent demonstration of intuitive (instinctive in the case of termites), spontaneous creation and self-organization. These intricately engineered masterpieces, created from the unplanned movement of many individual termites, are the tallest structures on earth relative to the size of their builders.

Although termites possess little individual physical capacity, they do have a strong sense of self and are constantly tuned into one another and their environment, which instinctively forms them into groups by attracting them to one another. They wander seemingly at will, bumping into one another and responding to the bump. When a critical mass of termites has gathered in a given location, their behavior shifts into an emerging action.

Their limited individual capabilities merge into a collective capacity, and they begin building their towers. A group on one side begins building an arch. Another group notices this and begins constructing the other side of the arch in a spontaneous action that meets in the middle sans engineer, boss, or planner.

When first studying these termite towers, entomologists sought a planner, a leader. But, after years of study, they concluded that the termite towers are examples of "emergent properties," meaning a group that is capable of behaviors simply unknowable when focusing on an isolated individual. Put differently, for all our study of the components of Nature, ourselves included, we will never see the potential for the collective possibility by evaluating an isolated piece.

There are many examples of emergent organizations, on both the micro- and macroscale, and they have some things in common. First, there is a very strong sense of self and purpose in each individual participating in the collective. Individuals know who they are collectively, what they want to be, or where they want to go (their vision); they are tuned into themselves and to one another. Second, they recognize the value of quality relationship and the necessity of a free flow of information within the group or system. Third, each individual may wander at will, bump into others with attractive energy, and respond, a behavior contrary to most traditionally organized interactions because it cannot be planned.

Each of us is capable of infinitely more than the sum of our parts, far exceeding our expertise in planning. Lists, calculations, or manipulations cannot access the quantum potential of our spontaneous creativity. It can only be accessed by a willingness to let something better than a rigid plan happen and participate intuitively in the happening. It is acting out of psychologically mature awareness, with full responsibility and participation in the dance of life.

To get the maximum benefit from any plan, people would do well to let creative spontaneity be part of the process. A plan must always be flexible and open to serendipitous opportunities, for which the exact outcome may be uncertain, but if the direction feels right intuitively, then *trust in the result* has to be the guiding principle.[18]

Policy

Laws and legal mandates contain inherently conflicting language regarding what may and may not be allowed in the name of planning, although the intent of the law is usually clear. But, agencies, either because of tradition or because of the instruction of a political administration, all too often use the interpretation of a specific policy to get around a given law and its mandates, even one with clear intent. Policy is thus used to meet corporate/political desires rather than the biophysical necessities of the social-environmental planning and the environmental considerations wherefor the law was originally intended.

Policy is often a seriously weak link within agencies because values cannot be legislated or mandated by law. So, despite the best intentions of public law, policy is used, by those with vested economic/political interests, to "legally" circumvent the "intent"—the spirit—of the law; an example is rezoning for economic reasons when the original zoning was for biophysical reasons to protect the *intergenerational quality* of community life.

Then, to fix the problems of such unethical abuses of policy, policy is used to justify rewarding or subsidizing various people to cause them to fulfill their legal duties as well as their moral obligations to the public, present and future. Such incentives are simply moral bribes.

To create and accept sound policies on environment and development, we must first agree that the long-term health of the environment takes precedence over the short-term profits to be made through careless or continual development. Then, we have to agree that biophysical sustainability is primarily an issue of managing *ourselves* by controlling our behavior and secondarily an issue of how we treat our environment. Thus, we come to a different kind of distinction about sustainability: Nothing can or will be sustained without our first deciding what we choose to sustain and develop and why and what we choose not to sustain or develop and why.

Shifting Nature's landscape toward a culturally oriented landscape requires a balance between those paths of development that are sustainable and those that are not. There are situations for which development is consistent with creating an enjoyable, productive, and sustainable, culturally oriented landscape. But, everything that is sustained or developed in a finite world is selectively chosen. Only in a constantly expanding world could we avoid the choices of what to sustain and develop and where, how, and why—or what not to sustain.

The path of development we choose is based on and controlled by policies, stated and unstated. Each policy is either a true or a false reflection of social values enshrined in public law; in that sense, the path of development may be more or less cooperative and environmentally benign or more or less competitive and environmentally malignant. But, whichever path we choose, that choice is ours. We cannot escape it. In other words, we need to create policies that safeguard, as much as possible, the flexibility of future choice.[19]

Leadership

Professor and author David Orr is of the opinion that we should expect more from our leaders than we do. "Never has the need for genuine leadership been greater, and seldom has it been less evident."[20] We need leaders, according to Orr, with the demonstrated humility of Czech President, Václav Havel:

> In time I have become a good deal less sure of myself, a good deal more humble. ... Every day I suffer more and more from stage fright; every day I am more afraid that I won't be up to the job. ... More and more often, I am afraid that I will fall woefully short of expectations, that I will somehow reveal my own lack of qualifications for the job, that despite my good faith I will make ever greater mistakes, that I will cease to be trustworthy and therefore lose the right to do what I do.[21]

In addition to the kind of leadership Orr espoused, long-term social-environmental planning requires a leader who encapsulates the aforementioned, second of Robert Rodale's elements of a big idea. Namely, they are generalists (systems thinkers) with a leadership advantage over specialists (product-oriented thinkers).[22] Moreover, a leader needs to understand what auto manufacturer Henry Ford did when he said, "Coming together is a beginning; keeping together is progress; working together is success."[23] In the end, it is the collective heart of the people that counts; without people, there is no need for leaders. Chinese philosopher Lao-Tzu thought a good leader was one who "when his work is done, his aim fulfilled, they will all say, 'We did this ourselves.'"[24] Such is servant leadership.

A good servant leader is a person with a fine balance between the masculine and feminine aspects of their personality. Such balance is critical because servant leadership often means putting relationships ahead of immediate achievement and knowing when each is important.

A servant leader intuitively knows that service is an attitude, not a function. Hence, such a leader does what is right from moral conviction, usually expressed as enthusiasm, causing people to want to follow with action.

A leader is one who values people and helps them transcend their fears so they might be able to act in a manner other than they are capable of doing on their own. This is the essence, the first rule, of true leadership. As such, it calls to mind a scene from the movie *The Karate Kid II*, in which Miyagi, a Japanese man who is the central character, is translating the rules of karate displayed on the walls of the Miyagi family dojo in Okinawa: "Rule number one: Karate is for defense only. Rule number two: First learn rule number one." In leadership this might translate as follows: Rule number one: Leadership is service to others based on inner strength of character. Rule number two: First learn rule number one.

A leader's power to inspire "followership" comes from a sense of inner authenticity because the individual has a vision that is other-centered rather than self-centered. Such a vision springs from strength, those universal principles that govern all life with justice and equity, as opposed to the relatively weak foundation of selfish desire. It is the authenticity to which people respond, and in responding they validate a leader's authority.

"Managerialship," on the other hand, is of the intellect and pays minute attention to detail, to the letter of the law, and to doing the thing "right" even if it is *not* the right thing to do. A manager relies on the external, intellectual promise of new techniques to solve problems and is concerned that all the procedural pieces are both in place and properly accounted for—hence the epithet "bean counter."

Good managers are thus placed at a disadvantage when put in positions of leadership because all such people can do is rise to their level of incompetence and remain there, in which case an ounce of image is worth a pound of performance. Similarly, a leader placed in the position of managerialship is equally inept because the two positions require vastly different skills.

Only an effective leader can guide the process of long-term social-environmental planning. An effective manager is the one who keeps the process running smoothly. Simply put, a leader knows what the right thing to do is, whereas the manger knows how to do the thing right.

A leader must be the servant of the parties involved. Servant leadership offers a unique mix of idealism and pragmatism. The idealism comes from having chosen to serve one another and some higher purpose, appealing to a deeply held belief in the dignity of all people and the democratic principle that a leader's power flows from commitment to the well-being of the people. Leaders do not inflict pain, although they must often help their followers to bear it in uncomfortable circumstances, like compromise. Such leadership is also practical, however, because it has been proven over and over that the only leader soldiers will reliably follow when risking their lives in battle is one they feel to be competent *and* personally committed to their safety.

A leader's first responsibility, therefore, is to help the participants examine their sense of reality, and the leader's last responsibility is to say "thank you." In between, one has to provide and maintain momentum as well as be effective. That said, most people confuse effectiveness with efficiency. *Effectiveness* is doing the right thing, whereas *efficiency*, in this sense, is doing the thing right to ensure effectiveness. Without this kind of leadership, no amount of policy (vision and goals) will suffice to guide long-term planning, no matter how well it is characterized.

Behavioral Constraints in Long-Range Social-Environmental Planning

To set the stage for this section, it is necessary to revisit the vision and goals of long-range planning.

Vision. A vision is the philosophical vanishing point on a distant horizon; it is the ideal that gives value to life's often-arduous journey. The ideal for long-term social-environmental planning is to pass forward a healthy, vibrant Earth that provides humanity with the best opportunity to achieve its highest potential in creating and maintaining social-environmental harmony.

Goals. A goal constitutes a clarification of the vision by elucidating a major area of accomplishment that needs to occur if the philosophical underpinning of the vision is to be upheld and realized. In this sense, goals represent the primary dimensions of the vision that must be monitored to determine the state of our environment's condition at any given time. It is vital that goals in long-term social-environmental planning include:

- The arrest and repair of environmental degradation by making small, timely corrections dictated by the best scientific data, thereby minimizing potential human-caused, catastrophic events. Such environmental protection and repair is to be accomplished by modifying long-term human behavior through education and legal mandates because the reversal of social-environmental problems is impossible by quick, technological fixes without substantive modification in personal and social behavior.
- The transformation of the current, exploitive socioeconomic system into one that is friendly—or at least benign—to both people and the environment. Such transformation is to be mediated by law or central-government planning and direction to deal with landscape-level sustainability. Any socioeconomic system that is developed must debit as well as credit to measure *and* show the "real" cost of doing business.

- The widely accepted understanding that women must have parity with men if the world's human population is to be controlled with any semblance of dignity and longevity. This endeavor will require a long and arduous educational process—and humility on the part of the male population.
- The consciousness of and accountability for the appropriateness of the technology that is created and introduced into the environment, which means conducting extensive risk analyses based first and foremost on Nature's biophysical constraints and last on economic cost-benefit analyses. When such cost-benefit analyses are done, they must include *real costs* to future generations based on a system of genuine economic indicators.

To achieve this vision and its attendant goals, the members of each generation have to accept a number of inherited behavioral constraints (how our decisions are interwoven among Nature's biophysical principles) that some people may find conceptually onerous. Nothing can be purposefully shifted in our social-environmental paradigm, however, without a commensurate shift in human behavior—the cause of the environmental effect. What is more, the discipline with which the members of each generation govern their own behavior will determine the nature and severity of the constraints they face during their tenure in leadership and life—as well the nature, number, and severity of the constraints each generation passes forward as its legacy to the next.

While the preceding paragraph sounds rather gloomy because it emphasizes constraints on human behavior, keep two things in mind. First, in a democracy, any law passed to limit the behavioral transgressions of the few is necessarily restrictive and thus affects all citizens the same whether or not they have in any way been complicit in abusing the rights of others. Second, it is the "intent"—the *spirit*—of such laws to protect each individual's birthright to such things as clean air, pure water, and the greatest degree of equality (within and between genders, within and among races and generations) that people can devise in the "human commons" of the present for the present *and* all future generations.

The purpose of any constraint is to balance the social-environmental ledger so that the next generation—and those that follow—will not begin in greater debt than the present generation. If each generation will adhere to a strategy that brings the social-environmental ledger into greater balance, each generation in its turn should find change easier to make and thereby reduce the number or severity of the potential constraints. But, if a generation, like the middle-aged generation of today, is unwilling to curb its material appetites, its legacy to all younger generations will be the continual impoverishment of their biophysical inheritance, for which inheritance means having to cope with what is left behind.

The foregoing brings to bear four working principles in long-range social-environmental planning: The first dictates that all who are involved in the planning process address the generations of the future as a concrete extension of the human community instead of viewing them as abstract individuals. This extension means visualizing yourself on the receiving end of your generation's behavior and decisions as well as the social-environmental effects you would inherit and have to live with due to your decisions. Put differently, before you make a decision, walk a mile in the shoes you would have the next generation wear.

Second, the language is framed in both possibilities *and* probabilities to the extent scientific knowledge allows. In this case, potential positive (as well as negative) outcomes are viewed with as much detachment as humanly possible in order to make the wisest and most rational decisions based on the cumulative welfare of all generations.

The third requires children from the first grade through high school to be included in the entire planning process and given unequivocal "First Amendment" rights. They are, after all, the future and have to live with the consequence of decisions for which they have not been given a voice. Although the children's voices in the planning process would meet some of the more immediate requirements in guiding the direction society needs to go, the really long-term affects of planning have to be addressed by using hypothetical scenarios to bridge the abyss of intergenerational communication beyond that possible with today's children. It would be necessary and wise in the hypothetical scenarios to build on the desires of the children to have adults be positive and psychologically mature and to represent basic human values. Besides, values tend to be somewhat different for each generation, differences that must be accounted for in social-environmental planning.

Fourth, all life entails risks, but an essential value of long-term social-environmental planning is to minimize, as much as humanly possible, passing forward unnecessary risks (whether existing or newly created) from one generation to the next because each risk passed forward progressively compounds the synergistic effects of all risks.

Benefits of Long-Term Social-Environmental Planning

In addition to all the goals, parameters, and legal requirements embedded in the planning process, it is fundamental that leaders endorse the concept of persons, including children, an endorsement that begins with recognition, understanding, and acceptance of people's diversity and intuition in their creative gifts, talents, and skills. These gifts, talents, and skills will also be needed to design lifestyles that are in sustainable harmony with the environment, which ultimately affects the sustainable harmony of the whole world.

When local people empower themselves to work together in tapping the utmost powers of the mind, intuition, and experience in developing their sustainable community, they will traditionally reap these benefits:

- A defined course of action (a vision) that helps ensure that the selected course has a good *probability* of success.
- A process that serves as the foundation whereon all community activities are based and, as such, must result in answers to what, where, when, why, and how actions are to be taken and who will conduct the actions for whom. If these questions are answered in a manner satisfactory to all, the *probability* of success is enhanced and the *probability* of a destructive conflict is forestalled, but if perchance conflict arises, it can usually be resolved within the context of sustainable community development.
- A well-conceived plan that allows those responsible to determine what they are responsible for and provides people with the opportunity to gain clear insight with respect to their specific tasks in relationship to the function of a community as a whole.
- A process that helps a particular group of people communicate to others that the group is thoughtful in what it is doing and has a good *probability* of accomplishing its stated purpose.
- A process that will aid in monitoring and evaluating the community and its achievements.
- A periodic evaluation of a group's progress toward meeting the vision, goals, and objectives identified in the plan, which is critical for evaluating whether it is providing the promised services to its customers and supporters. This step is essential for the health and growth of any community or organization within a community.
- A vehicle through which the collective, long-range (100+-year) vision of community citizens can be realized because long-term social-environmental planning seeks options and solutions to current and potential social-environmental problems in a way that helps people focus their energy on their vision, goals, and objectives, thereby helping their community to achieve maximum use of its human talents and financial resources.
- A process that helps people influence what the present is and the future might be for the benefit of both today's citizens and tomorrow's generations.

With the foregoing peek at long-term social-environmental planning, it is time to examine some of the issues that demand immediate attention if society is to survive the twenty-first century with any semblance of harmony and dignity. The issues I have selected are those I believe to be among the most important for humanity to address honestly and with a great deal of humility because our response will decide the potential options for many, if not all, generations to come.

Summation

With the ongoing shift toward social parity in the world, Chapter 12 opened a discourse on long-term social-environmental planning, which is a time-scale of planning that few seem disposed to acknowledge in light of today's socioeconomic priorities, such as globalization and immediate profits. Nevertheless, how we choose to behave today will either expand or limit the choices of all generations to come because we set in motion an ongoing ripple, the never-ending story of cause and effect. We have no choice in this because we exist. And in existing, we change the world for better or worse, depending on our attitudes and the decisions we make in accord with them. Any and all changes will revolve around four basic relationships: (1) intra-personal, (2) interpersonal, (3) between people and the environment, and (4) between people in the present and those of the future.

Finally, the relationship between people in the present and those of the future is a test of both the foresight (flexibility or rigidity) infused into the social-environmental harmony we create in our time and place in history and, consequentially, the ability of those future generations to respond to what we have done—our legacy.

Chapter 13 focuses on some of the challenges we need to face with respect to social-environmental planning. While this may sound onerous, we can successfully address each and every one of these challenges if we so choose. It is, after all, the choice of individuals taken in the collective of a democratic majority—the will of the people, as it were. The question before us is always the same: How will we choose—solely for our immediate benefit or for the common good of all generations?

Notes

1. Anatole France. http://thinkexist.com/quotes/anatole_france/ (accessed January 25, 2009).
2. The preceding discussion of ecopsychology is based on Mary-Jayne Rust. Seeking health in an ailing world. *Resurgence,* 224 (2004):16–18.
3. The foregoing discussion of relationships is based on Chris Maser. *Sustainable Community Development: Principles and Concepts.* St. Lucie Press, Delray Beach, FL (1997).
4. Elizabeth Sherrill. The power of a promise. *Guideposts* (August 1998) (page numbers not available).
5. Robert Rodale. "Big New Ideas—Where Are They Today?" Unpublished speech given at the Third National Science Technology Society (STS) Conference, February 5–7, 1988, Arlington, VA.

6. E. W. Sanderson, M. Jaiteh, M. A. Levy, and others. The human footprint and the last of the wild. *BioScience*, 52 (2002):891–904.
7. The foregoing discussion of a biological living trust is based on Chris Maser. *Our Forest Legacy: Today's Decisions, Tomorrow's Consequences*. Maisonneuve Press, Washington, DC (2005).
8. Edward Nickens. Saving the spirit trees. *American Forests*, 109 (2003):35–39.
9. Gary Lantz. Mapping the big trees. *American Forests*, 109 (2003):7–9.
10. Bruce E. Tonn. 500-Year planning: A speculative provocation. *Journal of the American Planning Association*, 52 (1986):185–193; Bruce E. Tonn. Philosophical aspects of 500-year planning. *Environment and Planning A*, 20 (1988):1507–1522.
11. Jay Walljasper. The speed trap. *Utne Reader* (March-April 1997):41–47.
12. Wolfgang Sachs. The speed merchants. *Resurgence*, 186 (1998):6–8.
13. Gregory Bateson. The Oak Beams of New College, Oxford. http://life-abundantly.blogspot.com/2007/10/oak-beams-of-new-college-oxford.html (accessed January 25, 2009).
14. Robert Luce. *Legislative Problems; Development, Status, and Trend of the Treatment and Exercise of Lawmaking Powers*. Da Capo Press, New York (1971).
15. William J. Quirk and R. Randall Bridwell. *Judicial Dictatorship*. Transaction Publishers, Edison, NJ (1995).
16. George Monbiot. Why we conform. *Resurgence*, 221 (2003):16–17.
17. M. K. Gandhi. *Non-Violent Resistance (Satyagraha)*. Courier Dover Publications, New York (2001).
18. The preceding discussion of termites, emergent properties, and planning is based on Kathy Gottberg. Confessions of an ex-planner. *The Quest*, 9 (1996):12–14.
19. The preceding discussion of policy is based on Chris Maser. Do we owe anything to the future? In: *Multiple Use and Sustained Yield: Changing Philosophies for Federal Land Management?* Proceedings and summary of a workshop convened on March 5 and 6, 1992, Washington, DC, 195–213. U.S. Government Printing Office, Washington, DC, 1992. Congressional Research Service, Library of Congress. Committee Print No. 11.
20. David W. Orr. The case for the earth. *Resurgence*, 219 (2003):16–19.
21. Ibid.
22. Robert Rodale. "Big New Ideas—Where Are They Today?" Unpublished speech given at the Third National Science Technology Society (STS) Conference, February 5–7, 1988, Arlington, VA.
23. Henry Ford. http://www.quotedb.com/quotes/2096 (accessed January 25, 2009).
24. John Heider. *The Tao of Leadership*. Bantam Books, New York (1988).

13

Challenges for Social-Environmental Planning

> We must face the prospect of changing our basic ways of living. This change will either be made on our own initiative in a planned way, or forced on us with chaos and suffering by the inexorable laws of nature.
>
> **President Jimmy Carter**[1]

Our sense of the world and our place in it is couched in terms of what we are sure we know and what we think we know. Our universities and laboratories are filled with searching minds, and our libraries are bulging with the fruits of our exploding knowledge. But, where is there an accounting of our ignorance?

Ignorance is not okay in our fast-moving world. We are chastised from the time we are infants until we die for not knowing an answer someone thinks we *should* know. If we do not have the correct answer, we can be labeled stupid, an appellation that is not the same as being ignorant about something. Being stupid is usually thought of as being mentally slow to grasp an idea, in other words—incompetent. Being judged ignorant, on the other hand, is to be chastised for not knowing the acceptable answer to a particular question.

Every question is a key that opens a door to a room filled with mirrors, each one a facet of the answer. Yet, only one answer is reflected in all of the mirrors in the room. If we want a new answer, we must ask a new question—open a door to a new room with a new key.

We, on the other hand, keep asking the same old questions—opening the same old door and looking at the same old reflections in the same old mirrors. We may polish the old mirrors and thereby hope to find a new and different meaning from the old answer to the same old question. Or, we might think we can pick a lock and steal a mirror from a new and different room with the hope of stumbling on a new, workable answer to the same old question.

Nevertheless, the old questions and the old answers have led us into the mess we are in today and are leading us toward the even greater mess we will be in tomorrow. In accordance with the above, we must look long and hard at where we are headed with respect to the quality of the world we leave as a legacy. Only when we are willing to risk asking really new questions can we find really new answers.

Heretofore, we have been more concerned with getting politically correct answers than we have been with asking biophysically intelligent questions. Politically correct answers validate our preconceived economic/political desires. Biophysically intelligent questions would lead us toward a future in which environmental options are left open so that generations to come can define their own ideas of a "quality environment" from an array of possibilities.

A good question—one that may be valid for a century or more—is a bridge of continuity among generations. We may develop a different answer every decade, but the answer does the only thing an answer can do: It brings a greater understanding of the question. An answer cannot exist without a question, so the answer depends on the question we ask, not on the information we derive as the illusion of having answered the question.

In the final analysis, the questions we ask guide the evolution of humanity and its society, and it is the questions we ask—not the answers we derive—that determine the options we bequeath the future. Answers are fleeting, here today and gone tomorrow, but a question may be valid throughout the earthly tenure of humanity. Questions are flexible and open ended, whereas answers are rigid, illusionary cul-de-sacs, so the future is a question to be defined by questions. Yet, even questions, if they are to serve us well, must arise with grace from the ethos of the heart.

Ethos, a Greek word meaning "character" or "tone," is best thought of as a set of guiding beliefs, but in phrasing this guiding direction, a distinction needs to be made between *ethos* and *policy*. Policy, written in explicit terms, can be in the form of an order—the letter of the law. Ethos, on the other hand, is implicit and includes a guiding set of human values that is understood but cannot be easily captured and conveyed in writing—the heart of the law based on a sense of spirituality. Today, instead of a clearly articulated ethos, we are engaged in a "runaway" money chase as if there were no tomorrow.

Yet, an ethos, and the spirituality that underpins it, can be translated into nurturing a behavior as operational policy should one wish to do so. That is why ignorance, where spirituality dwells, is the sacred land from which spirituality visits our intellect to inform us of the world beyond knowledge, the world of intuitive knowing—something religion cannot do because it is *taught* as sacrosanct "knowledge." Here, it would be wise to heed American author T. S. Eliot, who admonished: "Do not confuse information for knowledge, nor knowledge for wisdom." And, it is wisdom that is often found lacking in current bureaucratic decisions—a lack of wisdom that increasingly passes unconscionable, compounding risks to all generations.

To make this century one of healing the rift between forest and city, society and its environment, requires a beginning, and it seems that a logical place to begin is by deciding what we humans want from Earth. Once we know that, we have to figure out how we must behave to allow Earth to provide what we require. Whatever we decide will necessitate an examination of our collective philosophical foundation that is today badly outdated with respect to the biophysical sustainability of the direction in which society is now headed; because of that, let us scrap the notion of "carrying capacity" with respect to our human population and substitute "cultural capacity."

Cultural Capacity

The underpinnings of social values and chosen lifestyles are rooted in people's thoughts and values—their philosophy based on their cultural myths, translated into lifestyles. Moreover, it is the cultural underpinnings of their chosen lifestyles that ultimately affect the land they inhabit. *Lifestyle* is commonly defined as an internally consistent way of life or style of living that reflects the values and attitudes of an individual or culture, both materialistically and spiritually.

We in Western industrialized society have made lifestyle synonymous with "standard of living," which we practice as a search for ever-increasing material prosperity. That notwithstanding, to have a viable, sustainable environment as we know it and value it necessitates reaching beyond the strictly material to see lifestyle as a sense of inner wholeness and harmony derived by living in such a way that the spiritual, environmental, and material aspects of our lives are in balance with the capacity of the land to produce the necessities for that lifestyle.

Whether a given lifestyle is even possible depends on the *biological carrying capacity* of an area, which is the number of animals, including people, that can live in and use a particular landscape without impairing its ability to function in an ecologically specific way. If we want human society to survive the twenty-first century in any sort of dignified manner, we need the humility to view our own population in terms of local, regional, national, and global carrying capacities because the quality of life declines in direct proportion to the degree to which the habitat is overpopulated. In this sense, carrying capacity has no built-in margin of safety from environmental degradation due to overpopulation because overpopulation is the transgression of a biophysical threshold, which represents the culmination of cumulative effects that accrued throughout a lag period in the invisible present.

Is there, one might wonder, a higher moral purpose that will be served by cramming eight, nine, or even ten billion people into a world where space and the other necessities of life are finite—and shrinking? I say *shrinking* for two reasons: First, people are living longer and so using more resources than heretofore; second, each new person brought forth who exceeds the balance of those who die needs—and ideally is given—an equal share of life's necessities. But, this *equal share* means the amount of land required to meet each person's necessities is continually shrinking on a per capita basis.

Because the quality of human life and the maximum numerical carrying capacity of people on Earth *cannot be maximized simultaneously*, we are required to consider an alternative. Could we substitute the idea of cultural capacity for biological carrying capacity and have a workable proposition for society? Cultural capacity is a chosen quality of life that has the best chance of being sustainable without endangering the environment's productive potential. For example, the more materially oriented the desired lifestyle of an individual or a society, the more resources that lifestyle requires, the smaller the human population must be per unit area of landscape. Cultural capacity, then, is a purposefully defined balance between how we want to live, the real quality of our lifestyles and of our society, and how many people an area can support in that lifestyle on a sustainable basis. Cultural capacity of any area will be less than its carrying capacity in the biological sense and must be maintained through voluntary reproductive discipline, both personal and collective.[2]

We can predetermine local and regional cultural capacity and adjust our population growth accordingly, such as the small network of land trusts, government agencies, and foundations that have purchased thousands of acres of coastline in California since 2000 to protect its 1,100-mile coast from any more development and thus keep it as open space for the enjoyment of all generations.[3] If we choose not to balance our desires with the land's capabilities, the depletion of the land will determine the quality of our cultural experience and our lifestyles. So far, we have chosen not to balance our desires with the capabilities of the land because we have equated "desire," "need," and "demand" as synonyms with every itch of "want." In so doing, we have lost sight of biophysical reality.

To maintain a predetermined lifestyle dictates that we ask new questions:

1. How much of any given resource is necessary for us to use if we are to live in the lifestyle of our choice?
2. How much of any given resource is necessary to leave intact as a reinvestment of biological capital in the health and continued productivity of the ecosystem?
3. Do sufficient resources remain, after biological reinvestment, to support our lifestyles of choice, *or* must we modify our proposed lifestyles to meet what the land is capable of sustaining?

These questions are vital because a life "necessity" is a very different proposition from the collective, egotistical "desire, want, need, demand" syndrome—so arguments about the proper cultural capacity revolve around what we think we want in a materialistic/spiritual sense, as well as around what the land can produce in a biophysically sustainable sense. Cultural capacity is a conservative concept given finite resources and well-defined values. By first determining what we want in terms of lifestyle, we may be able to determine if Earth can support our desired lifestyle and so direct how we must behave with respect to the environment if we are to maintain our chosen lifestyle. To make the concept of cultural capacity more concrete, let us examine the decline of the Hawaiian paradise where many unanticipated, unwanted changes will last forever.

Until people found the Hawaiian Islands, perhaps one new species evolved every 10,000 years. This number is significant because the Hawaiian Islands surpass even the Galapagos Islands, off the coast of Ecuador, in the number of species that evolved from a single ancestor. In Hawaii, at least fifty species have evolved from a common ancestor.

Beginning in the 1700s, the islands became a crossroads for Pacific travel, and early seafarers introduced domestic pigs, goats, horses, and cattle onto the islands as sources of fresh meat. Even within recent years, the introduction of foreign species of plants and animals has increased dramatically.

In addition to the obvious introductions of domestic animals, less-expected imports affected the islands. Bird malaria and bird pox, both of which are carried by mosquitoes, have had a severe impact on indigenous Hawaiian birds. Brown tree snakes, which have devastated the indigenous species of birds on Guam, have been intercepted on flights to Hawaii at least six times.

The banana poka, a passionflower vine, which is kept in check in its native South America by the feeding of insects, has no such controlling mechanisms in the islands. Consequently, since its arrival it has smothered 70,000 acres of forests on two islands and is threatening larger tracts.

By the mid-1990s, nearly two-thirds of Hawaii's original forest cover had been lost, including half the vital rain forests. As of November 1991, 90 percent of the lowland plains, once forested, had been destroyed. Of 140 species of native birds, only 70 remained, and 33 of those were in danger of extinction. Eleven more species were beyond recovery, and 37 species of plants indigenous to Hawaii were listed as federally endangered. By 1997, there were 152 more species proposed for federal listing. Ninety-three species of trees, shrubs, vines, herbs, and ferns were among the state's rarest plants, each having only about a hundred known surviving individuals as of 1995. At least five species had been reduced to a single individual. The cause of the decline is twofold: (1) the cumulative effect of people's careless, unplanned, unbalanced conversion of the land from Nature's design to society's cultural design in the form of agriculture, ranching, and residential use and (2) the introduction of nonindigenous species of plants, insects, and mammals.

The results include the loss of the forests that once intercepted and generated rainfall and protected the coral reefs and beaches from siltation caused by the erosion of soil. Forest loss, coupled with the extinction of indigenous plants and animals, affects every level of the islands' economy and cultural heritage, such as the generation of unique materials for clothing, textiles, ornaments, canoes, and scientific study. Because its cultural capacity has been grievously exceeded, Hawaii has become largely a paradise lost.[4]

The Hawaiian experience illustrates clearly and dramatically that it is vitally important for humanity, especially in Western industrialized societies, to understand and accept that biological *and* genetic diversity are the underpinnings of functional diversity, which in turn is the foundation of a sustainable quality of life. The more we strip the world of its diversity in the pursuit of money, the more we degrade our potential lifestyles. It would behoove us, therefore, to get serious about saving diversity by embracing the notion of cultural capacity while there is still time—because time is running out due to our exploding human population.

Controlling Our Human Population

The world's human population will only be controlled within viable parameters when women are granted gender equality, control over their bodies, and the right to choose the number of children they have and the age at which to have them. Population control, in comparison to birth control, is a relatively recent concept in our contemporary thinking. Thomas Malthus, the political economist, addressed this subject in 1798 because he was concerned about the yawning abyss between the rich and the poor in the British Isles. Although Malthus wrote that the human population would at some point begin to overtake its available supply of food, he decried the use of contraceptives and was condemned as an alarmist. Nevertheless, he understood that if humanity did not control its own population, Nature would—and in ways most unpleasant. After all, we humans added three times our number to the earth's surface in the twentieth century and in doing so put more human pressure on its life support systems than had existed in all prior recorded time.[5] Was Malthus right after all?

In 1900, the total population of the United States, including immigrants (both legal and illegal), was 75 million. The population was 100 million in 1915, it was 200 million in 1968 and in 2006 was 300 million.[6] It increased by 32.7 million during the last decade because of a large wave of young immigrants and a rate of birth that has overtaken the rate of death. This was the largest increase ever experienced in the population in a single decade. Extrapolations indicate this number may increase to around 400 million by 2050 and 571 million in 2100. Immigrants and their offspring, who came here after the year 2000, will supply two-thirds of that growth.

Unfortunately, there are people who still think the United States is so large that it has endless open space available for development—and thus more people. Indeed, when I fly across the country and survey the land from the window of an airplane, the uninhabited areas do seem to stretch forever into the distance. But, how much of that land is arable and hospitable to farming or human habitation? How much of it has potable water readily available in a sustainable supply?

Looking at the global human population, through statistics produced by the United Nations, does indeed give me pause—1.6 billion people worldwide in the year 1900, with 6.1 billion in 2002 as I finished writing my book *The Perpetual Consequences of Fear and Violence*, 8.9 billion by 2050, and in 2100 perhaps around 10 billion people inhabiting the globe.[7] Of course, no one is able to estimate the number accurately, but this seems like an acceptable prediction.

The population in the nonindustrialized countries is increasing by seventy-five million people each year and is predicted to rise about 73 percent in the next fifty years. This would add up to 87 percent of the global population by 2050. How is society going to manage such growth? What will it do to human migration patterns throughout the world? If wars continue to plague society, where will the refugees go? With the ever-destructive military mentality, which both incites and participates in wars that deplete natural resources, how will nations, overwhelmed by refugees, care for them?

On top of this, the aging population of the nonindustrialized nations will increase almost twice as fast as those of the industrialized countries, in part because people are living longer. By 2020, people over age 80 will increase by 22 million in the industrialized part of the world, whereas people over age 80 will increase by nearly 52 million in the nonindustrialized part of the world during the same time.

Since the 1960s, the longevity of women in western Asia and Africa has increased by seven to ten years, and by the year 2020, that longevity may increase by another five to eight years. The number of women over age sixty was estimated to be 208 million in 1985. By 2020, it is projected to be 604 million, with 70 percent of them living in poor, nonindustrialized countries, where most will be daily worrying about the source of their next meal.

Clearly, growth in the human population is confronting all of us with dilemmas of the most profound nature. We face such problems as the destruction of forests, degradation of the land, rivalries over access to potable water, destruction of the oceans, depletion of an increasing number of natural resources, inability to maintain levels of production, and outbreaks of disease. Burundi, in Africa, for instance, has a population of six million, one-third suffering from malaria. In Botswana and South Africa, a large percentage of the population has AIDS but no access to medicines because of the exceedingly high cost of pharmaceuticals.

In the United States alone, without these medicines and improved knowledge about maintaining good health, people would have died much younger, and the population would be around 140 million rather than the 281 to 283 million it is today. When these improvements spread worldwide, the high rate of death in such places as Africa and the Indian subcontinent declined significantly, and the population boomed. Despite the extant problems of disease in many African nations, by 2050 the population as a whole is expected to increase from eight hundred million to two billion.[8]

If poor countries develop tastes for and can afford some of the industrialized nation's lifestyles, humanity is going to outstrip the available supply of goods and resources in a very short time, such as (1) potable water, the competition for which will become the cause of wars[9]; (2) food and the arable land to grow it because the acreage per individual shrinks each time one more person is born than dies—and even faster when urban sprawl and huge dams are taken into account; and (3) energy, such as wood from forest and fossil fuels. With respect to forests, they are being summarily cut down for short-term monetary gains in countries that can least afford to lose them and the free ecological services they perform.[10] With respect to fossil fuels, the question becomes one of how much pollution the environment can tolerate before it crosses the irreversible threshold of degradation that humanity cannot well survive.

Such an outcome is not necessary, of course, but to avoid it, *men* must raise the level of their consciousness with respect to the way they treat women, such as:

- giving women the choice of how many children to have and when, as well as *valuing them beyond* their abilities to satisfy a male's sexual urges and to bear children—especially sons
- providing women with *safe access* to legal abortions and adequate counseling about birth control
- *ending genital mutilation*
- seeing that gender *equality is given*—and is taught—throughout the entire educational cycle
- instilling the fact that, while men and women differ biologically in some ways, *they are equally human*, meaning women deserve equality with men in all aspects of social life, including educational, social, political, and economic
- giving women *positions of political authority* because women focus on relationships beyond violence and power mongering

Many changes are required in restructuring society as a whole so that women are equal partners in the human experiment of cohabitation. We the people—and men in particular—ignore the equality of women at our collective peril.

Another part of the answer resides in the earliest grades in school, where girls and boys need to be taught the importance of gender equality and the shared responsibility for limiting the size of the world's human population. It remains to be seen whether the world's men—*especially* those who control the world's organized religions with such dogmatic male attitudes—are up to the task.

To rectify the problem of male domination, we must make women the *subjects* of—*not the objects* of—policies dealing with population by insisting they occupy at least half of all managerial and policy positions in the area of family health and planning, as well as human population dynamics and stabilization, and the environment. To make women subjects, as opposed to objects, will demand change in the attitudes of most men—worldwide—toward girls and women. It is also critical to the current population crisis that women have the necessary authority to:

- correct the inadequacies of women's health, including reproductive health
- address the inequitable distribution of food, water, and shelter
- ensure access to safe, legal, abortion services
- end genital mutilation
- stop the slave trade in girls and women as sex objects
- accede to demands for gender equality
- empower women in the ecological, economic, social, and political arenas

If we men are at all serious about curbing the world's looming overpopulation, we, in both secular and religious life, will openly and honestly give women equality in all opportunities. We will stop telling them what they can and cannot do with their bodies that, after all, are entrusted to their consciences—*not ours.* We will allow women to choose when to have children and how many to have. And, men will find the courage to accept *and do* our part in controlling the human population *by having vasectomies.*[11] There is much riding on a positive outcome, not the least of which is human survival with any semblance of quality and dignity, a notion that brings forward the specter of pollution because of what we introduce into the environment.

What We Introduce into the Environment

While we can manage what we introduce into the environment by the choices we make, once something is introduced, it is inexorably out of our hands and our control despite the fact that we assume certain biophysical principles to be constant values in the economic sense. By *certain biophysical principles,* I mean the depth and fertility of the soil where a forest grows, the quality and quantity of the water falling on the forest, the quality of the air infusing the forest, the quality of the sunlight reaching the forest, the biological and genetic diversity driving the forest, and the climate where the forest lives.

Because it is assumed biophysical principles can be converted into constant values in the economic sense, it is also assumed that forests are infinite suppliers of the goods and services for the long-term sustainability of cities. Since we assume these principles to be economic, constant values in both quality and quantity, they are omitted from our economic and planning models and, even more basically, from our thinking. Yet, each is a biophysical variable.

Soil, for instance, is eroded in two ways, chemically and physically, and we are doing both. Water and air are polluted with chemicals. Air pollution has a direct effect on the entire world and in turn affects the quality and quantity of the sunlight that energizes all of Earth's ecosystems—including the oceans. Through our linear, product-oriented, economic thinking, we are simplifying these same ecosystems and causing untold extinctions in the process. And, today's changes in global climate are affecting everything else. *Nothing in Nature is static.* Nature gives us only variables that are constantly changing and thus irreversible.

If we disarm the world of all weapons of war and continue to pollute and kill the global ecosystem that sustains us, our survival becomes only a matter of time. If we escape to another planet and do not change our thinking, again the only question for humanity is time because we take ourselves with us wherever we go.

We can, of course, consciously change our behavior if we so choose simply by managing the only thing we can manage—*ourselves.* Clearly, we do not have to pollute as we do. We can choose otherwise.

Think about the refuse generated by our throwaway society. How many garbage dumps, euphemistically called "sanitary landfills," are there now, and how many more will be needed by the end of this century?

Although much of what is discarded by our society may break down or be consumed by microorganisms over varying amounts of time, there is much that will undoubtedly be around for centuries. As the world's population continues to grow and towns and cities expand ever farther into the countryside, the human-generated garbage will increase proportionately while the available land area in which to hide it will continually shrink. This situation is untenable in the design and sustainability of our social-environmental aspirations.

Author and businessman Paul Hawken put it succinctly:

> Industry has transformed civilization and created material wealth for many people. But it only succeeds by generating massive amounts of waste. Our production systems function by taking resources, changing them to products and discarding the detritus back into the environment. This process is overwhelming the capacity of the environment to metabolize our waste, and, as a result, the health of our living systems is slowly grinding down. There are simply too many manufacturers making too much too fast. ... By the time you get to the consumer, 98 percent of the problem has already occurred.[12]

We must therefore specifically ask: How much human-generated waste can we convert into food for microorganisms so that it can continually cycle throughout the environment as a renewable source of energy? In addition, while it is still in the design stage, we must ask of each new item that technology proposes to produce: Will this item be biodegradable?

If the answer is "yes," then we must ask: By what mechanism will it be biodegradable? To what degree will it be biodegradable? In what time frame will it be biodegradable? And, under what conditions is it biodegradable?

If the answer is "no," then we must ask: Why is it not biodegradable? Can it be made to be biodegradable? If not, is there a biodegradable substitute that can serve the same purpose? If not, can one be made? If not, *do not make it.*

According to Paul Hawken, there is a growing movement (the Factor Ten Club) in both Europe and the United States to radically reduce our requirements from resources. The Factor Ten Club consists of scientists, economists, and experts in public policy who are calling for a 90 percent reduction (factor ten) in the use of materials and energy over the coming forty to fifty years.[13]

The goal of factor ten is to invent products, technologies, and systems that deliver the services people want without 90 percent of the fuel, metal, wood, packaging, and waste that is currently used. Keep in mind, admonished Hawken, that we want *the service* a product provides, *not the thing itself.* For example, we want clean clothes not necessarily the detergent that cleans them.[14]

The upshot is that society can no longer afford to produce and use things that are wasteful and are not rapidly biodegradable. But, unless these questions are specifically asked, new technology will most likely produce more and more things that are nonbiodegradable to increasing degrees. This does not mean that planned obsolescence as a corporate marketing strategy will disappear; it only means that whatever is discarded will take up more and more space in an unusable form for ever-longer periods of time, like industrial chemicals, metals like aluminum, some plastics, and nuclear waste already do.

Nonbiodegradable materials, especially those that are toxic and capable of spreading, can make planning for social-environmental sustainability exceedingly difficult, even at the scale of a local community. The negative effects of such materials are often more poignant at the bioregional scale, however, where they may affect an entire supply of water.

And, speaking of nonbiodegradable materials, what about those of past decades that now lie buried in sanitary landfills? Is any of it reusable? Yes, according to the Berkshire County Council, in England, which is going to "mine" old landfills for reusable glass, metals, and plastics to become raw materials for the recycling business. This action (along with increasing present-day recycling and reduction of such things as flagrant overpackaging) will begin to clean the environment, create jobs, and increase the life span of already-established landfills.[15] With this in mind, let us approach renewable energy.

Renewable Energy

We clearly need to focus on developing, using, and shifting entirely to clean, renewable sources of energy that, in the scale of humanity, can be thought of as infinite. I am speaking of power from the wind and the sun. Although many people may be in favor of wind power or solar power, they often do not want solar panels on house roofs in their neighborhood because of aesthetics.[16]

Nevertheless, these forms of energy can be harnessed and used without damaging the earth's biophysical systems, but we will have to wean ourselves from oil, an emancipation that people with vested interests in the oil and gas business resist mightily. Eventually, even hydroelectric power will prove to be too damaging environmentally for its continued use. This said, power from both wind and sun require open space.

Open Space

Open space, like water, is available in a fixed amount. But unlike water, open space is visibly disappearing at an exponential rate. Once gone, it is gone—unless, of course, rural communities, and perhaps even cities, are torn down to reclaim it. The commitment to maintain a matrix of open spaces within and surrounding a community is critical to both its sustainability and its economic viability, especially a small community in a nonurban setting. Although there are multiple reasons why a community might want to save open spaces, the protection of local water catchments is a crucial one because, as previously mentioned, water is a nonsubstitutable requirement of life for which there is a finite storage capacity. Its availability throughout the year will determine both the quality of life in a community and consequently the value of real estate. Accordingly, it behooves a community to take every possible measure to maximize and stabilize both the quality and quantity of its *local* supply of water.

Water

By local supply, I mean water catchments in the local area under local control, as opposed to water catchments in the local area under the control of an absentee owner with no vested interest in the community's supply of water. Such absentee ownership could be a person, corporation, government body, or agency beyond local jurisdiction. Various absentee owners are increasingly taking over control of the world's supply of freshwater, which they deem to be a "human need," and so a "commodity," instead of a birthright as part of the global commons.[17]

With this in mind, it is wise to purchase as much of the local water catchments as possible and maintain them as open space expressly for the purpose of capturing water and storing it in the ground, where it can be purified as it flows slowly toward the wells it recharges. This will help prevent those people with wells from needing municipal water and so will help to maintain a more predictable demand—and supply—over time. And, those people with wells, who do not pay for municipal water, could be charged a fee for using water from the community-owned water catchment as a means of helping defray the costs of maintaining the catchment's health.

If outright purchase of a water catchment is not possible, a community could conceivably enter into a long-term lease or contract to rent the catchment, with control over what is done on it. Then, it might be possible to accrue monthly or annual payments toward the price of purchasing the land at a later date. Such an arrangement could benefit the owner in terms of a steady income at reasonable tax rates while allowing some long-term sustainable use of the land.

Another alternative might be a tax credit payable to the landowner if the community could work in conjunction with the owner to protect the water catchment's inherent value to the community itself. There probably are other options, but the important option is to secure the purchase of local water catchments in community ownership as part of the open space program to maintain and protect the quality of life and the local value of real estate. An added value may be that some part of a water catchment could also be used as seasonal, nondestructive recreational space for the community as a whole—no destructive off-road vehicles or motorbikes.

Communal Space

Open space for communal use is central to the notion of community and is rapidly becoming a premium of a community's continued livability and the stability of the value of its real estate. Of course, continual economic growth, at the expense of open space, will profit a few people in the present, but it will ultimately steal from everyone in the future.

Yet, for communal open space to have maximum value over time, the community must have a clear vision of what it wants so that the following questions can be answered in a responsible and accountable way:

1. What parcels of land are wanted for the communal system of open space?
2. Why are they wanted? What is their functional value: capture and storage of water, habitat for native plants and animals, local educational opportunities, recreation, aesthetics, or all of these values?
3. How much land is necessary to fulfill the first two needs.

4. Can one project the value added to the quality of life or the future value of real estate, including that outside the community's urban growth boundary?

Surrounding Landscape

The land surrounding a community's municipal limits gives the community its contextual setting, its ambiance, if you will. The wise acquisition of open spaces in the various components of the surrounding landscape, whether Nature's ecosystem or that of culture, protects, to some extent at least, the uniqueness of the community's setting and hence the uniqueness of the community itself. And, the value added, both spiritual and economic, will accrue as the years pass.

Agricultural Cropland

According to C. J. De Loach, "The objective of agriculture is to encourage the growth of a foreign organism, a crop, at a high density and to suppress ... organisms that might compete with it."[18] Yet, it was not always so cut and dried, as noted by David Pimentel: "When man dug holes here and there and planted a few seeds for his food, ample diversity of species remained, but this resulted in small crop yields both because of competition from other plants [weeds] and because insects, birds, and mammals all took their share of the crop."[19]

Be that as it may, in modern agricultural practice in North America, large fields are often planted with a single species. This specialization has resulted from an ever-expanding, centralized, corporate power base, aided by technology in an increasingly mechanized society. In the process of centralizing corporate power, a greatly simplified, increasingly fragile, labor-intensive, energy-intensive environment has been created through the following changes in the land:

1. Increased specialization of farms (growing fewer crops in larger fields) caused amalgamation of small, individual fields.
2. Increased size of individual farms due to specialized, corporate farms replaced small, diversified family farms.
3. Increased use of modern machinery can more easily and more economically be operated in large, single-crop fields.
4. Increased clearing of fencerows was done to gain more land for agriculture, where one mile of fencerow may occupy one-half acre.[20]
5. Increased use of large sprinkler irrigation systems eliminated uncultivated irrigation ditches and their banks and mined underground water sources.

6. Many uncultivated earthen banks of irrigation ditches were replaced with concrete.
7. The elimination of nectar corridors for indigenous pollinators is causing a drastic decline in their numbers and consequently their ability to pollinate crops.
8. There is constant human control of crops with fungicides, herbicides, insecticides, rodenticides, or all four if the desired production is to be forthcoming.
9. Federal aid has been given to farmers through the Agricultural Stabilization and Conservation Service for various types of land "reclamation"—a misnomer.

As these factors reduced the habitat for many species of wild plants and animals, they also increased the tendency for these same plants and animals, living in areas surrounding the croplands, to be perceived as exerting a constant negative influence on production—even those that perform ecosystem services, such as plants that feed wild pollinators. When wild species, especially animals, use agricultural crops as habitat, they are normally termed *pests*. Whether a species is a pest is a matter of perception based on some level of competitive tolerance, which wanes rapidly when money is concerned.

Small, diversified family farms were excellent habitat for wildlife. They provided increased structural diversity and so increased habitat diversity through a good mix of food, cover, water, and mini-open spaces within surrounding, otherwise rather homogeneous, croplands.

The many small, irregular fields with a variety of crops created an abundance of structurally diverse edges, and tillage offered a variety of soil textures for burrowing animals. Uncultivated fencerows and ditch banks provided strips that acted as primary habitat for species, such as insectivorous songbirds and wild pollinators, in addition to providing travel lanes among the fields for other species.

Replacement of small family farms by large corporate ones, dependent on mechanization and specialized monocultural crops, caused a drastic decline in wildlife habitats within and adjacent to croplands. Because of the decreased crop stability—increased crop vulnerability—resulting from the greatly simplified "agricultural ecosystem," farmers are more and more inclined to view wild or nonagricultural plants and animals as actual or potential pests to their crops.

In addition to stripping habitats from fencerows surrounding fields to maximize tillable soil and get rid of unwanted plants and animals, modern agriculture is killing the soil and poisoning the groundwater, ditches, streams, rivers, estuaries, and ultimately the oceans of the world with chemicals, which is clearly neither biologically nor culturally sustainable. How can such destructive agriculture be redeemed?

Meeting with the local farmers and discussing the kinds of produce that could be grown to make the community as self-sufficient as possible can redeem destructive agriculture. The economic viability of the remaining small family farmers can be ensured by loyally purchasing their produce. Organically grown produce may cost a little more, but it is healthier, and organic farming heals the soil and does not add polluting chemicals to the water.

Any community can purchase open space in the form of fencerows—even around the edges of cemeteries, sewage treatment ponds, and city reservoirs—along which to allow fencerow habitat to re-create itself. Then, in addition to minihabitats in and of themselves, the few uncultivated yards could once again act as longitudinal corridors for the passage of wildlife from one area to another as well as a feeding station for indigenous pollinators. Living fencerows would also make the landscape more interesting, more appealing to the human eye, and add again the songs of birds and the colors of flowers, butterflies, and leaves to the passing seasons.

The point is to find out what worked sustainably in the past and begin re-creating it in the present, and, if problems arise, work together to resolve them. The only way to create, maintain, and pass forward the sense of community is by working together because the friendliness of a community is founded on the quality of its interpersonal relationships, of which small, family farmers ideally need to remain an integral part.

Forest Land

If a community is in a forest setting, the forest more likely than not is a major contributor to the community's image of itself; in addition, it may comprise an important water catchment. Furthermore, if the community is, or has been, a "timber town," then most of the forest may well have been converted to economic tree farms; thus, maintaining an area of native forest may be of even greater value. And, if some old trees are included in the area, its spiritual value may well be heightened and its value as habitat for some plants and animals greatly enhanced.

On the other hand, if what surrounds a community is no longer forest but rather an economic tree farm, a purchased area could be helped to evolve again toward a forest. As such, its aesthetic and spiritual values would increase, as would its potential educational value. Comparing a relatively sterile tree farm to a real forest can teach us much.[21] One will find, for instance, that a forest harbors a far greater diversity of plants and animals than does a tree farm, even one near the age of cutting, especially within riparian areas.

Riparian Areas

Riparian areas can be identified by the presence of vegetation that requires free or unbound water and conditions more moist than normal. These areas may vary considerably in size and the complexity of their vegetative cover because of the many combinations that can be created between the source of water and the physical characteristics of the site. Such characteristics include gradient, aspect of slope, topography, soil, type of stream bottom, quantity and quality of the water, duration of flow, elevation, and the kind of plant community.

Riparian areas have the following things in common:

1. They create well-defined habitats within much drier surrounding areas.
2. They make up a minor portion of the overall area.
3. They are generally more productive than the remainder of the area in terms of the biomass of plants and animals.
4. Wildlife use riparian areas disproportionately more than any other type of habitat.
5. They are a critical source of diversity within an ecosystem.

There are many reasons why riparian areas are so important to wildlife, but not all can be attributed to every area. Each combination of the source of water and the attributes of the site must be considered separately because:

1. The presence of water lends importance to the area since habitat for wildlife is composed of food, cover, water, space, and privacy. Riparian areas offer one of these critical components and often all five.
2. The greater availability of water to plants, frequently in combination with deeper soils, increases the production of plant biomass and provides a suitable site for plants that are limited elsewhere by inadequate water. The combination of these factors leads to increased diversity (composition) in the species of plants, a more complex structure, and thus richer functional dynamics of the biotic community.
3. The dramatic contrast between the complex of plants in the riparian area with that of the general surrounding vegetation of the upland forest or grassland adds to the structural diversity of the area. For example, the bank of a stream that is lined with deciduous shrubs and trees provides an edge of stark contrast when bordered by coniferous forest, grassland, agricultural fields, or even suburbia. Moreover, a riparian area dominated by deciduous vegetation provides one kind of habitat in the summer in full leaf and another type in winter after leaf fall.
4. The shape of many riparian areas, particularly the linear nature of streams and rivers, maximizes the development of edge effect that is so productive in terms of riparian-oriented wildlife.

5. Riparian areas, especially those in coniferous forests, frequently produce more edges within a small area than would otherwise be expected based solely on the structure of the plant communities. In addition, many strata of vegetation are exposed simultaneously in stair-step fashion. This stair-stepping of vegetation of contrasting form (deciduous versus coniferous or otherwise evergreen shrubs and trees) provides diverse opportunities for feeding and nesting, especially for birds and bats.
6. The microclimate in riparian areas is different from that of the neighboring area because of increased humidity, a higher rate of transpiration (loss of water) from the vegetation, more shade, and increased movement in the air. Some species of animals are particularly attracted to this microclimate.
7. Riparian areas along intermittent and permanent streams and rivers provide routes of migration for wildlife, such as birds, bats, deer, and elk. Deer and elk frequently use these areas as corridors of travel between high-elevation summer ranges and low-elevation winter ranges. These areas also offer cover while traveling across otherwise open ground.[22]

In addition, riparian areas supply organic material in the form of leaves and twigs that become an important component of the aquatic food chain. They also supply large woody debris in the form of fallen trees, which in turn form a critical part of the land-water interface, the stability of banks along streams and rivers, and instream habitat for a complex of aquatic plants as well as aquatic invertebrate and vertebrate organisms. Moreover, riparian areas close to coastal shores supply critical large driftwood to the oceanic ecosystem.[23]

Setting aside riparian areas as undeveloped open space means saving the most diverse, and often the most heavily used, habitat for wildlife in proximity to a community. Moreover, people are strongly drawn to riparian areas for recreation and sites on which to build their homes. Riparian areas are also an important source of large woody debris for the stream or river whose banks they protect from erosion.[24] Further, these areas are periodically flooded in winter, which, along with floodplains, is how a stream or river dissipates part of its energy. Such dissipation is an important function. Without it, floodwaters would cause significantly more damage than they already do in settled areas.

Floodplains

A *floodplain* is a flat area of land that borders a stream or river and is subject to flooding. Like riparian areas, floodplains are critical to maintain as open areas because, as the name implies, they frequently flood. These are areas where storm-swollen streams and rivers spread out, decentralizing the velocity of their flow by encountering friction caused by the increased surface area of their temporary bottoms, both of which dissipate much of the floodwater's energy.

It is wise to include floodplains within the matrix of open spaces for several other reasons:

1. They will inevitably flood and put any human development at risk, regardless of efforts to steal the floodplain from the stream or river for human use (witness the Mississippi River).
2. They are critical winter habitat for fish.[25]
3. They form important habitat in spring, summer, and autumn for a number of invertebrate and vertebrate wildlife that frequent the water's edge.[26]
4. They can have important recreational value. Try to steal land from a stream or river, and sooner or later it will reclaim it, at least temporarily, and at great cost to the thieves. In this case, "centralized generalization" can be of help.

Centralized Generalization

There are many places, primarily abroad, so conceived and built that people can fulfill their basic daily requirements within walking distance of their homes or bike lanes for cyclists; either way, they allow people to get exercise and keep the air cleaner by leaving their cars parked. The convenience of *centralized generalization* of goods and services in mixed-use areas allows people to meet one another on the street and in the shops, where they get to know one another and often stop to exchange pleasantries. There are also places, like a particular cafe or soda fountain, a particular shade tree, or a community well, where local people daily gather and visit; in Taos, New Mexico, it is the local post office.

As people get to know one another, they became familiar with what each person does and on whom they can rely when in need of help. Because people know one another, they look out for one another in a free exchange of mutual caring. Of course, there are social problems and interpersonal conflicts, but mutual well-being is a great incentive to settle them peacefully.

Some people in these places do not even own an automobile and do not need one. I realize that such a notion, here in the United States, is tantamount to heresy, but I experienced the lack of such ownership while working in Slovakia in the 1990s. It was marvelous since you could go wherever you wanted on mass transit. And, it may someday be a necessary condition of the future if the biophysical *quality* of human life is to be maintained within sufficient limits to pass forward to future generations.

The Urban-Wildlife Interface

The final challenge is the urban-wildlife interface, which unchecked urban sprawl is greatly exacerbating. Even in my hometown of Corvallis, Oregon, mountain lions are hunting deer within the city limits. One or two have been within eyesight of primary schools. Let one child be killed by a mountain lion, and the big cats will pay a terrible price for our human failure to take their life's requirements into account as we design our communities to continually encroach on their dwindling habitat, from local villages to big cities. And, this is just one species. This problem is serious, compounding, and the time to deal with it is *now*.

Summation

Chapter 13 highlighted a few of the social-environmental challenges that need to be confronted, accepted, and overcome to make this century one of healing the rift between Everycity and Everyforest, the ripples of which would spread benefits worldwide for all generations.

Chapter 14 is titled "Where Leaders Dare to Go" because it will require the kind of leadership that focuses steadfastly on a vision for a sustainable future based on sound, farsighted, other-centered, social-environmental planning than is today common in the world. The kind of leadership I am speaking of is one devoid of ego, that rare person who is willing to share leadership when the opportunity arises.[27]

Notes

1. Richard Heinberg. *The Party's Over: Oil, War and the Fate of Industrial Societies* (rev. ed. 2). New Society Publishers, Gabriola Island, BC, Canada (2005).
2. Garrett Hardin. Cultural carrying capacity: A biological approach to human problems. *BioScience*, 36 (1986):599–606.
3. Jim Wasserman. California save coast from growth. *Albany (OR) Democrat-Herald, Corvallis (OR) Gazette-Times* (March 21, 2004).

4. The discussion of the loss of biological diversity in Hawaii is based on Daniel B. Wood. Report details decline of Hawaiian paradise. *The Oregonian* (November 7, 1991), which is a joint report that took a decade for the United States Fish and Wildlife Service, the Hawaii Department of Land and Natural Resources, and The Nature Conservancy of Hawaii to prepare; Rocky Baker. Mending fences: Lessons in island biodiversity protection from Hawaii. *East-West Center Working Papers: Environmental Series,* No. 45 (1995):1–46.
5. Niles Eldredge. Will Malthus be right? *Time Magazine* (November 8, 1999): 102–103.
6. U.S. population to surpass 300 million. *Science News,* 170 (2006):238.
7. If you want a more in-depth discussion of our overpopulation, which is largely a result of how girls and women are treated in today's world, see Chris Maser. *The Perpetual Consequences of Fear and Violence: Rethinking the Future* Maisonneuve Press, Washington, DC (2004). Further, this section on population is adapted from that publication.
8. The discussion of population numbers is based on Jean H. Lee. World population boom expected. *Corvallis Gazette Times* (February 28, 2001); Eric Schmitt. U.S. population has biggest 10-year rise ever. *The New York Times* (April 3, 2001); Robert Weller. Western states no longer dependent on California for growth. *Corvallis Gazette-Times* (July 10, 1999); Carol Savonen. Population growth: A blessing or a curse? In: *Looking for Oregon's Future.* Oregon State University Extension Service, Covallis, OR, 2001, p. 5; Jim Rydingsword. Longevity, for its own sake, is not enough. *Corvallis Gazette-Times* (March 25, 2002).
9. Michael Ruane. Fickle water controls all of us. *Albany (OR) Democrat-Herald, Corvallis (OR) Gazette-Times* (August 15, 1999); Fred Bridgland. Looming water wars. *Sunday Herald* [Glasgow, Scotland]. *World Press Review* (September 19, 1999).
10. Chris Maser. *Sustainable Forestry: Philosophy, Science, and Economics.* St. Lucie Press, Delray Beach, FL (1994); Janet N. Abramovitz. Learning to value nature's free services. *The Futurist,* 31 (1997):39–42.
11. The discussion of gender equality is based on Maser. *The Perpetual Consequences.*
12. Paul Hawken. Undoing the damage. *Vegetarian Times* (September 1996):73–79.
13. Ibid.
14. Ibid.
15. Lorna Howarth. Wealth in waste. *Resurgence,* 180 (1997):23.
16. Jim Carlton. People favor solar power—but not in their neighborhood. *The Wall Street Journal* (February 25, 2004).
17. Maude Barlow. Water democracy. *Resurgence,* 219 (2003):30–32; Vandana Shiva. Captive water. *Resurgence,* 219 (2003):33–35.
18. C. J. De Loach. The effect of habitat diversity on predation. *Proceedings Tall Timber Conference on Ecological Animal Control by Habitat Management,* 2 (1971):223–241.
19. David Pimentel. Population control in crop systems: Monocultures and plant spatial patterns. *Proceedings Tall Timber Conference on Ecological Animal Control by Habitat Management,* 2 (1971):209–220.
20. N. W. Moore, M. D. Hooper, and B. N. K. Davis. Hedges. I. Introduction and reconnaissance studies. *Journal of Applied Ecology,* 4 (1967):201–220.
21. Chris Maser. *Forest Primeval: The Natural History of an Ancient Forest.* Sierra Club Books, San Francisco (1989); Maser. *Sustainable Forestry.*

22. The discussion of riparian zones is based on Jack Ward Thomas, Chris Maser, and Jon E. Rodiek. Riparian zones. In: *Wildlife Habitats in Managed Forests—The Blue Mountains of Oregon and Washington,* Jack W. Thomas (tech. ed.), 40–47. U.S. Government Printing Office, Washington, DC, 1979. USDA Forest Service, Agricultural Handbook No. 553.
23. Chris Maser and James R. Sedell. *From the Forest to the Sea: The Ecology of Wood in Streams, Rivers, Estuaries, and Oceans*. St. Lucie Press, Delray Beach, FL (1994).
24. Ibid.
25. Ibid.
26. Ibid.
27. A discourse on leadership is beyond the scope of this book. For such a discussion, see Chris Maser. *Vision and Leadership in Sustainable Development*. Lewis Publishers, Boca Raton, FL (1998).

14

Where Leaders Dare to Go

> Most people look at the world as it is, and ask, "Why?"
> I look at the world as it could be, and ask, "Why not?"
>
> **George Bernard Shaw[1]**

This last chapter outlines some of the places leaders must dare to go if social-environmental sustainability is to be the mainstay of the twenty-first century. And, this journey of leaders begins with a question and ends with a question.

Do we owe anything to the future? If so, we must understand and accept that there are no external fixes for internal, moral imperatives; there are only internal shifts of consciousness, morally correct intentions, and biophysically appropriate behaviors. We must also understand and accept that all we can bequeath to the generations of the future are options—the right to choose as we have done. To protect that right of choice, we must ask new, morally sensitive, biophysically imperative, future-oriented questions—questions that determine the quality of lifestyle we want to have and that we want our children and grandchildren to be able to have.

But, first and foremost, we need to determine how much of a given, biophysical resource is necessary to leave intact in the environment as a biological reinvestment in the health and continued productivity of the ecosystem. We must, at any cost, be it economic or political, protect the quality of the air, water, and soil of our home planet if humanity and its society is to survive. It is also critical for us to view the environment from the standpoint of biological and cultural necessities as opposed to limitless, personal wants, desires, needs, and demands, and, if necessary, *alter our lifestyles* to reflect what the global ecosystem can in fact sustainably support.

Our continued survival as a society dictates that we account for the intrinsic, biophysical value of all natural resources as well as for their conversion potential into money. In this context, we have to accept that the long-term health of the environment takes precedence over the short-term profits to be made through exploitation and continual development. Concurrently, we must convert our society—immediately, rapidly, consciously, and unconditionally—to a version of capitalism that views long-term, ecological wholeness and biological richness of the environment as *the measure of long-term economic health* through the employment of genuine economic indicators.

To this end, it is absolutely necessary that we pass clearly stated, precisely worded, unambiguous laws wherein the intent is so simply stated that it cannot be hidden by distorted, bureaucratic policy. It is thus essential that we create social-environmental policy that is commensurate with biophysical sustainability and cultural capacity. Further, it is vital that such policy simultaneously protects these values from the negative, irreversible aspects of continual development based solely on garnering ever-greater economic bottom lines.

We must also accept that the only sustainability for which we can "caretake" is one that ensures the ability of an ecosystem to adapt to a changing global climate, and that means caretaking for choice, *maximum biophysical diversity*, regardless of the perceived short-term economic and political costs. In turn, biophysical diversity can be protected only by caretaking for a long-term *biophysically sustainable* condition on the landscape and by abandoning our cherished, unworkable notion of immediate, *"sustained," ever-increasing yield* of raw materials for industrial purposes. To achieve such a condition, we need to stop today's practice of "managing for habitat fragmentation" by focusing only on commodity-producing resources. We must instead focus on caretaking for the connectivity of habitats to help ensure the ecological wholeness and the biological richness of the patterns we create across landscapes.

If we are to be successful trustees of the future's right of choice, we must unfailingly "manage" the only thing we really can manage—ourselves, including what we introduce into the environment—in such a way that we conscientiously live within the biophysical-moral confines of our cultural capacity. The importance of living within our cultural capacity cannot be overemphasized because the great and only gift we can give to our children, our grandchildren, and beyond is the right of choice and something of value from which to choose.

Finally, we need hope, as Czech President Václav Havel said in 1990:

> Either we have Hope within us or we do not. ... It is a dimension of the soul and is not essentially dependent on some particular observation of the world. Hope is an orientation of the spirit, an orientation of the heart. It transcends the world that is immediately experienced and is anchored somewhere beyond the horizons. Hope in this deep and powerful sense is not the same as joy [or optimism] that things are going well or willingness to invest in enterprises that are obviously headed for early success, but rather an ability to work for something because it is good, not just because it stands a chance of succeeding. Hope is definitely not the same thing as optimism. It is not the conviction that something will turn out well, but the certainty that something makes sense regardless of how it turns out. It is Hope, above all, which gives the strength to live and continually try new things.[2]

"We the people" are the trustees of the future's options. Our challenge, therefore, is to find and test our moral courage and political will. To succeed requires the body politic to act in the following manner if we are to maintain Everyforest as a biological living trust and Everycity as a cultural living trust,

both being part of the human commons among generations—the birthright of *every man, woman, and child*:

The Questions We Ask

1. Ask new, responsible, biophysically and culturally relevant questions. With respect to new questions, those that will raise the level of social consciousness about the future of humanity, I have found an extremely insightful article—one that can go a long way in securing a dignified future for all generations. The article is titled "The Identification of 100 Ecological Questions of High Policy Relevance in the UK."[3] Despite their confinement to the United Kingdom, these questions are far more than merely ecological. Indeed, they address the very heart and soul of social-environmental planning and sustainability. What is more, they are eminently pertinent to all people everywhere—without exception—despite the geographical limitation of the title. Moreover, striving to answer them will raise the collective level of our consciousness above that responsible for the problems in the first place. Although the questions are framed in terms of the United Kingdom, specifically England, they can easily be adapted to any geographical area worldwide. As well, additional questions of bioregional significance can be added if necessary.

These questions fall into fourteen categories listed here, with one question from each. Although I have left the categories in the order they were listed, I have done my best to select the questions most pertinent to social-environmental planning and, if necessary, edited the question to make it more generic:

1. Ecosystem services: What are the benefits of protected areas in terms of water resources, carbon sequestration, and other goods and services relative to nonprotected land?
2. Farming: How do current agricultural practices affect the conservation value and extent of nonagricultural habitats, such as woodland edges, remnants of prairies, hedgerows, and aquatic habitats (including ditches), and how can detrimental impacts be mitigated?
3. Forestry: What overall acreage (hectares), area configuration, age structure, and spatial distribution of trees is necessary for the long-term survival of species dependent on old trees and forested habitats?
4. Fisheries, aquaculture, and marine conservation: What are the ecological impacts on the marine ecosystem from fecal matter, pesticides, and undigested food that escapes from aquaculture?

5. Recreation and field sports: What are the impacts of recreational activities on biodiversity?
6. Urban development: How can social-environmental planning be used to create the maximum benefit for wildlife within existing and new urban development, such as street corridors, urban parks and other open spaces, and those areas between urban and rural settings?
7. Alien and invasive species: What criteria should be used to determine when to intervene to deal with invasive species?
8. Pollution: Of those chemicals currently or potentially released into the environment that (individually or in combination) are now, or are likely to become, significant environmental problems, what will these problems be, and how is it best to deal with them?
9. Climate change: How can we increase the resilience of habitats and species to cope with climate change?
10. Energy generation and carbon management: What are the consequences of biofuel production for biodiversity at field, landscape, and regional levels?
11. Conservation strategies: With what precision can we predict the ecological impact of different policy options and the ecological effects of social-environmental planning?
12. Habitat protection (management) and repair: What are the costs and benefits of concentrating the protection and maintenance of habitats on designated sites in comparison with spreading efforts across the wider countryside?
13. Connectivity and landscape structure: How can social-environmental planning be used to protect landscape-level habitat mosaics for the conservation of diverse taxa that operate on different spatial scales?
14. Making space for water: How can social-environmental planning be used to assist in flood control through ecologically appropriate habitat protection and repair, and how would such planning affect biodiversity?[4]

The questions we ask today about how to achieve social-environmental sustainability, the level of consciousness with which we derive the answers, and the courage (personal and political) to act in accord with our decisions will design the legacy we leave all generations. The time is *now*. It is all we have—or ever will.

So, here is perhaps the most important question to ask now, right now: I'm just one person: What difference can I make? "If you think you are too small to make a difference," said author Anita Roddick, "you have never been in bed with a mosquito."[5]

The Lifestyle We Choose

1. Determine the quality of lifestyle we want and that we want our children to be able to have.
2. Determine how much of any given resource is *necessary* for us to use if we are to live in the lifestyle of our choice—and, given our collective decision, determine what we must do to manage our local and regional population accordingly.
3. Determine how much of any given resource must necessarily be left intact as a reinvestment of biological capital in the health and continued productivity of every ecosystem to safeguard its biophysical wealth for all generations.
4. Compare the necessities of the land with the necessities of our desired lifestyle.
5. Determine how we need to behave toward the environment to help ensure that our chosen lifestyle can be maintained without stealing options from, squandering the inheritance of those who follow.
6. Determine how we must adjust our lifestyle to meet what the land is biophysically capable of sustaining *if* it cannot support our preferred lifestyle.
7. Accept that biophysical sustainability is *primarily* an issue of *managing ourselves* in terms of our behavior, our population size, our chosen lifestyle, and only *secondarily* an issue of caretaking our environment.
8. Decide in terms of social-environmental sustainability what to sustain, develop, and why and what not to sustain, develop, and why.

The Economics We Employ

1. Account for the intrinsic, biophysical value of all natural resources—*not just* for their conversion potential into money.
2. Accept that the long-term, biophysical health of the environment takes precedence over the short-term profits to be made through commercial exploitation.

3. Convert our society—immediately, consciously, and unconditionally—to a version of capitalism that views long-term ecological wholeness and biological richness of the landscape as *the measure of economic health.*
4. "It *is* possible [after all] to give a new direction to technological development … back to the real needs of man, and … to the *actual size of man*. Man is small, and, therefore, small is beautiful," as stated in 1973 by economist Fritz Schumacher, a consummate systems-thinker.[6]
5. Create policy that protects social-environmental sustainability from negative, irreversible aspects of nonsustainable, linear economic development.

The Laws and Policies We Enact

1. Pass clearly stated, precisely worded, unambiguous laws wherein the intent is so simply spelled out that it cannot be obfuscated by internal, bureaucratic policy.
2. Create social-environmental policy that is commensurate with social-environmental sustainability.
3. Create policy that protects social-environmental sustainability from negative, irreversible aspects of nonsustainable development.

The Landscape Patterns We Create

1. Accept that the only sustainability we can caretake is that which ensures the ability of an ecosystem to adapt to environmental change.
2. Accept that biophysical diversity can be protected only by caretaking an ecosystem within the biophysical constraints of a clearly crafted vision for social-environmental sustainability and by *abandoning* our cherished, unworkable notion of *sustained, ever-increasing yield of raw materials for industrial purposes.*
3. Caretake for spatial and temporal connectivity of habitats and stop today's practice of managing for fragmentation of the landscape by focusing solely on commodity-producing resources and resource areas.

We already have most of the laws and mandates necessary to give us license to caretake our environment in a biophysically sound manner, one that complies with the discussion here. We must now find the moral courage and the political will to follow both the intent *and the spirit* of those laws for the long-term good of all generations—regardless of the short-term economic costs and the political uncertainties. If current laws are not morally or biophysically sound, better ones can be passed as necessary. The choice is ours—a choice of well-being or survival, a choice ultimately determined by the discernment of human consciousness and the courage of leaders who dare to lead for the sake of this magnificent planet we call home. To the young generations of today and all the generations of the future, we bequeath the consequences. May we choose wisely.

Notes

1. George Bernard Shaw. The Danish Peace Academy. http://www.fredsakademiet.dk/library/avery/utopia.htm (accessed January 27, 2009).
2. Rory Spowers. Web of hope. *Resurgence,* 219 (2003):28–29.
3. William J. Sutherland, Susan Armstrong-Brown, Paul R. Armsworth, and others. The Identification of 100 ecological questions of high policy relevance in the UK. *Journal of Applied Ecology,* 43 (2006):617–627.
4. Ibid.
5. Anita Roddick. http://www.mini-iq.co.uk/launch/launch_01b.html (accessed January 27, 2009).
6. E. F. Schumacher. *Small Is Beautiful: Economics as If People Mattered.* Hartley and Marks Publishers Inc., Point Roberts, WA (1999).

Glossary

carbohydrate: Any of a group of chemical compounds, including sugars, starches, and cellulose, containing carbon, hydrogen, and oxygen only.

carbon: A naturally abundant nonmetallic element that occurs in many inorganic and all organic compounds.

carbon dioxide: A colorless, odorless, incombustible gas, CO_2, formed during respiration, combustion, and the decomposition of organic material.

chlorophyll: Any of a group of related green pigments found in organisms that posses photosynthesis.

chloroplast: A microscopic, ellipsoidal organelle in a green plant cell (see "organelle").

chroma (n.), chromatic (adj.): The quality of a color combining hue and saturation.

clay: Very fine-grained sediment that becomes plastic and acts like a lubricant when wet. Clay consists primarily of hydrated silicates of aluminum and is widely used in making bricks, tiles, and pottery.

clear-cutting: The act of cutting down and removing all the trees from a forested area.

Commons: That part of the world and universe that is every person's "birthright." There are two kinds of commons. Some are gifts of Nature, such as clean air, pure water, fertile soil, a rainbow, northern lights, a beautiful sunset, or a tree growing in the middle of a village; others are the collective product of human creativity, such as the town well from which everyone draws water or a museum of fine art.

Commons Usufruct Law: The personal right, in common with everyone else, to enjoy all the advantages derivable from the use of something held in common, provided the substance of the thing being used is not injured in any way.

disequilibrium: A point at which a system is out of balance.

Divine Principle: In the Christian sense, anything created by the One God, The Almighty.

endpoint: Point of termination.

Everycity: A generalization depicting the common features of cities.

Everyforest: A generalization depicting the common features of forests.

First Law of Thermodynamics: The total amount of energy in the universe is constant, although it can be transformed from one form to another.

Five Dynasties and Ten Kingdoms period: AD 907–979, one of the most tumultuous eras in Chinese history, called the Five Dynasties and Ten Kingdoms period, during which time five dynasties rose and fell within a few decades, and China fractured into several independent nation-states.

foraminifera: Minute marine organisms with exquisitely designed shells.

fusion: The merging or blending of two or more things or substances.

hiding cover: Topographical or vegetational cover that an animal uses for hiding in the face of potential danger.

hue: A specific shade of a particular color.

Law of Cosmic Unification: Functionally derived from the synergistic effect of three universal laws: the first law of thermodynamics, the second law of thermodynamics, and the law of maximum entropy production.

Law of Maximum Entropy Production: A system will select the path or assemblage of paths out of available paths that minimizes the potential or maximizes the entropy at the fastest rate given the existing constraints.

Levant: Former name of that region of the eastern Mediterranean that encompasses modern-day Lebanon, Israel, and parts of Syria and Turkey.

limestone: Sedimentary rock formed from the calcium carbonate skeletons and shells of marine organisms.

molecule: The smallest particle of a substance that retains its chemical and physical properties and is composed of two or more atoms bound together by chemical forces.

Nature: The forces and processes that produce and control all the phenomena of the material world.

old-growth: A forest that is past full maturity; the last stage in forest succession; a forest with two or more levels of canopy, heart rot, and other signs of obvious physiological deterioration.

organelle: A differentiated structure within a cell that performs a specific function.

photoreceptor: Cells specialized to sense or receive light.

photosynthesis: The process by which chlorophyll-containing cells in green plants convert incident light to chemical energy and synthesize organic compounds from inorganic compounds, especially carbohydrates from carbon dioxide and water, with the simultaneous release of oxygen.

predator: Any animal that kills and feeds on other animals.

pyroclastic flow: A turbulent mixture of hot gas and fragments of rock, such as pumice, that is violently ejected from a fissure and moves with great speed down the side of a volcano. *Pyroclastic* is Greek for "fire-broken."

quark: The fundamental unit of matter.

red algae: Marine algae in which the chlorophyll is masked by a red or purplish pigment.

rhodophytes: A general term for red algae (seaweed) from the Greek *rhodon*, "rose" and *phuton*, "plant."

sandstone: Sedimentary rock composed of particles of sand, primarily quartz, bound together with a mineral cement.

saturation: Chromatic purity, freedom from dilution with white.

Second Law of Thermodynamics: The amount of energy in forms available to do useful work can only diminish over time. The loss of available energy to perform certain tasks thus represents a diminishing capacity to maintain order at a certain level of manifestation (say a tree), and so increases disorder or entropy.

siltstone: A form of fine-grained stone composed of compressed silt.

stalactite: The most familiar ice-cycle-shaped structure found hanging from the ceilings of limestone caves. They grow by the separation of calcium carbonate from within a thin film of fluid flowing down their surfaces.

stalagmite: Mineral-rich waters dripping from the cave's ceiling onto its floor year after year forms the stalagmite (a mirror image of a stalactite).

stratosphere: The second major layer in the Earth's atmosphere, which is stratified in that warmer layers are above cooler layers.

succession: Progressive changes in species composition and forest community structure caused by natural processes over time.

thermal cover: Topographical or vegetational cover that an animal uses to ameliorate effects of weather.

usufruct (n.), usufructuary (adj.): The legal right for a person to use and derive profit from property belonging to someone else, provided that the property itself is not injured in any way.

wildlife: Animals, especially vertebrates, living in a natural setting.

Appendix: Common and Scientific Names of Plants and Animals

Plants	
Algae	
Red alga	*Polysiphonia* spp.
Trees	
Bigleaf maple	*Acer macrophyllum*
Douglas-fir	*Pseudotsuga menziesii*
Eucalyptus	*Eucalyptus* spp.
Giant sequoia	*Sequoia gigantea*
Great Basin bristlecone pine	*Pinus longaeva*
Norway spruce	*Picea abies*
Oregon ash	*Fraxinus latifolia*
Oregon white oak	*Quercus garryana*
Pacific madrone	*Arbutus menziesii*
Red alder	*Alnus rubra*
Red spruce	*Picea rubens*
Ponderosa pine	*Pinus ponderosa*
Western hemlock	*Tsuga heterophylla*
Western red cedar	*Thuja plicata*
Invertebrates	
Foraminifera	
Foraminifera	Foraminifera
Insects	
Ants	Formicidae
Vertebrates	
Fish	
Dusky farmerfish	*Stegastes nigricans*
Birds	
Acorn woodpecker	*Melanerpes formicivorus*
Cliff swallows	*Petrochelidon pyrrhonota*
Common flicker	*Colaptes auratus*
Cooper hawk	*Accipter cooperii*
Eagle owl	*Bubo bubo*
Golden eagle	*Aquila chrysaetos*

Great horned owl	*Bubo virginianus*
Green woodpecker	*Picus viridis*
Osprey	*Pandion haliaetus*
Pileated woodpecker	*Dryocopus pileatus*
Raven	*Covus corax*
Sand grouse	*Pterocles* spp.
Sand martin	*Riparia riparia*
Sparrow hawk	*Accipiter minullus*
Swainson's thrush	*Catharus ustulatus*
Swifts	Apodidae
Treecreeper	*Certhia familiaris*
Tree swallow	*Tachycineta bicolor*
Varied thrush	*Ixoreus naevius*
Wilson's Warbler	*Wilsonia pusilla*
Winter wren	*Troglodytes troglodytes*
White-headed woodpecker	*Picoides albolarvatus*
Mammals	
Bank vole	*Clethrionomys glareolus*
Barren-ground caribou	*Rangifer tarandus groenlandicus*
Bats	Chiroptera
Big horn sheep	*Ovis canadensis*
Bobcat	*Felis rufus*
Brush rabbit	*Sylvilagus bachmani*
Bushy-tailed woodrat	*Neotoma cinerea*
California red-back vole	*Clethrionomys californicus*
Common brushtail possum	*Trichosurus vulpecula*
Deer mouse	*Peromyscus maniculatus*
European red squirrel	*Sciurus vulgaris*
Fisher	*Martes pennanti*
Golden jackal	*Canis aureus*
Gophers	Geomyidae
Grizzly bear	*Ursus arctos*
Himalayan pika	*Ochotona himalayana*
Jungle cat	*Felis chaus*
Lesser bamboo rat	*Cannomys badius*
Lion	*Panthera leo*
Long-footed potoroo	*Potorous longipes*
Marten	*Martes americana*
Moles	Talpidae
Moose	*Alces alces*
Mountain beaver	*Aplodontia rufa*
Mountain lion	*Felis concolor*
Naked mole rat	*Heterocephalus glaber*
North American elk	*Cervus elaphus*

Otter	*Lutra* spp.
Sable	*Martes zibellina*
Shrews	Soricidae
Snow leopard	*Uncia uncia*
Snowshoe hare	*Lepus americanus*
Spotted skunk	*Spilogale putorius*
Tiger	*Panthera tigris*
Water vole	*Microtus richardsoni*
Weasels	*Mustela* spp.
Wildebeest	*Connochaetes* spp.
Wolf	*Canis lupus*
Wolverine	*Gulo gulo*

Index

M

Q

R